THE
MÖBIUS
STRIP

WORKS BY CLIFFORD A. PICKOVER

The Alien IQ Test
Black Holes: A Traveler's Guide
Calculus and Pizza
Chaos and Fractals
Chaos in Wonderland
Computers, Pattern, Chaos, and Beauty
Computers and the Imagination
Cryptorunes: Codes and Secret Writing
Dreaming the Future
Egg Drop Soup
Future Health
Fractal Horizons: The Future Use of Fractals
Frontiers of Scientific Visualization
The Girl Who Gave Birth to Rabbits
Keys to Infinity
Liquid Earth
The Lobotomy Club
The Loom of God
The Mathematics of Oz
Mazes for the Mind: Computers and the Unexpected
Mind-Bending Visual Puzzles (calendars and card sets)
The Paradox of God and the Science of Omniscience
A Passion for Mathematics
The Pattern Book: Fractals, Art, and Nature
The Science of Aliens
Sex, Drugs, Einstein, and Elves
Spider Legs (with Piers Anthony)
Spiral Symmetry (with Istvan Hargittai)
Strange Brains and Genius
Sushi Never Sleeps
The Stars of Heaven
Surfing Through Hyperspace
Time: A Traveler's Guide
Visions of the Future
Visualizing Biological Information
Wonders of Numbers
The Zen of Magic Squares, Circles, and Stars

THE MÖBIUS STRIP

DR. AUGUST MÖBIUS'S MARVELOUS BAND
IN MATHEMATICS, GAMES, LITERATURE,
ART, TECHNOLOGY, AND COSMOLOGY

CLIFFORD A. PICKOVER

THUNDER'S MOUTH PRESS
NEW YORK

THE MÖBIUS STRIP
Dr. August Möbius's Marvelous Band in Mathematics, Games, Literature, Art, Technology, and Cosmology

Published by
Thunder's Mouth Press
An Imprint of Avalon Publishing Group, Inc.
245 West 17th St., 11th Floor
New York, NY 10011-5300

Library of Congress Cataloging-in-Publication Data is available.

ISBN-10: 1-56025-952-3
ISBN-13: 978-1-56025-952-7

9 8 7 6 5 4 3 2 1

Book design by Maria Elias
Printed in the United States of America
Distributed by Publishers Group West

Möbius was the epitome of the absentminded professor. He was shy and unsociable . . . and so absorbed in his thought that he was forced to work out a whole system of mnemonic rules . . . so as not to forget his keys or his inseparable umbrella. . . .

What was perhaps his most impressive discovery—that of one-sided surfaces such as the famous Möbius strip—was made when he was almost seventy, and all the works found among his papers after his death show the same excellence of form and profundity of thought.

—Isaak Moiseevich Yaglom, *Felix Klein and Sophus Lie*

There's a theory that the universe is forever folding back and over on itself like a cross between a Möbius curve and a wave. If we catch that wave, it will be quite a ride.

—*Gene Roddenberry's Andromeda,* "Answers Given to Questions Never Asked," Episode 401

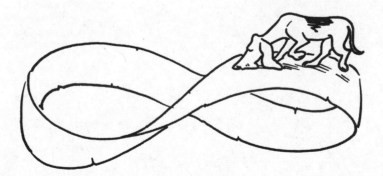

CONTENTS

3.

A Brief History of Möbius the Man

In which we encounter Möbius's family tree, simultaneity in science, Schulpforta, Paul Julius Möbius, Möbius syndrome, Johann Benedict Listing, the "king with five sons" problem, Möbius's mathematical output, Karl August Möbius, the false dawn animal, the Möbius maze puzzle, and Möbius licentiousness. . . .

4.

Technology, Toys, Molecules, and Patents

In which we encounter the Rhennius machine, Roger Zelazny's *Doorways in the Sand*, Möbius patents and toys, Möbius molecules, mathematics patents, lemniscates, astroids, Reuleaux triangle drill bits, conveyor belts with twists, surgical retractors, Möbius electrical components and train tracks, knot patents, the metaphysics of shoelaces, chirality, Lipitor, Paxil, Zoloft, Nexium, thalidomide, Advil, enantiomers, *Methanobacterium thermoautotrophicum*, Möbius plant proteins that induce labor in African women, Möbius crystals, the Noah's ark puzzle, and Möbius strips in fashion and hairstyle. . . .

5.

Strange Adventures in Topology and Beyond

In which we encounter Benoit Mandelbrot, fractals, parameterizations, a conical helix, butterfly curves, paradromic rings, Leonhard Euler, Antoine-Jean Lhuillier, chromatic numbers, projective planes, the four-color theorem, "The Island of Five Colors," Möbius's triangulated band, Johann Listing, homeomorphisms, ghosts, the fourth dimension, Immanuel Kant, Johann Carl Friedrich Zöllner, Henry Slade, Alfred Schofield's "Another World," turning spheres and doughnuts inside out, optiverses, the Boy surface, cross-caps, Roman surfaces, the fantastic Möbius function, the Mertens conjecture, the Riemann zeta function, Möbius palindromes, the amazing π, coprimality, graph theory, hexaflexagons, Möbius shorts, Möbius tetrahedra, Möbius triangles, solenoids, Alexander's horned sphere, prismatic doughnuts, perfect square dissections, the squiggle map coloring puzzle, the cannibal torus, the pyramid puzzle, and Möbius in pop culture. . . .

A Few Final Words

In which we encounter Stanislaw Ulam, Franz Reuleaux, Georg Bernhard Riemann, Zen koans, *Eternal Sunshine of the Spotless Mind*, Harlan Brothers, Marjorie Rice, Roger Penrose, Arthur C. Clarke, the Mandelbrot set, an ambiguous ring, and Möbius strips in business and government.

ACKNOWLEDGMENTS

The frontispieces for each chapter include figures from U.S. patents, which are described in chapter 4. All of these patents prominently feature the Möbius strip. I discussed some of the ideas in this book at my "Pickover Think Tank," located on the Web at:

http://groups.yahoo.com/group/CliffordPickover/

I thank group members for the wonderful discussions and comments. I also thank sculptor John Robinson and math professor Ronnie Brown for supplying the image of themselves next to John's trefoil knot sculpture in figure 2.7. Visit their Web sites at www.popmath.org.uk, www.JohnRobinson.com, and www.BradshawFoundation.com for more information. Belgian computer artist and mathematician Jos Leys (www.JosLeys.com) provided computer graphics renditions of knots and one-sided surfaces. Other graphics contributors include Andrew Lipson, M. Oskar van Deventer, Cameron Browne, Nicky Stephens, Christiane Dietrich-Buchecker, Jean-Pierre Sauvage, Rob Scharein, Tom Longtin, Henry S. Rzepa, David Walba, Dave Phillips (www.ebrainygames.com), George Bain, Teja Krasek, Rinus Roelofs, and Donald E. Simanek. To create the puzzle piece tessellations in figures 7.31–7.33, Tom Longtin used Jonathan Shewchuk's Triangle program (www.cs.cmu.edu/~quake/triangle.html).

I thank Dennis Gordon, Nick Hobson, Kirk Jensen, George Hart, Mark Nandor, and Graham Cleverley for useful comments and suggestions, and I thank Brian Mansfield (www.brianmansfield.com) for his wonderful "cartoon" drawings that I use throughout this book. April Pedersen drew the picture of the dog walking on a Möbius strip, used on the quotation page near the front of this book.

For an excellent introduction to August Ferdinand Möbius, see *Möbius and His Band: Mathematics and Astronomy in Nineteenth-Century Germany* edited by John Fauvel, Raymond Flood, and Robin Wilson. This book also describes how nineteenth-century German mathematicians and astronomers developed into the most powerful and influential thinkers in the world.

Martin Gardner's *Mathematical Magic Show* and numerous other books by Gardner provide delightful introductions to the Möbius band and topology. Many Web sites provide useful information on the Möbius strip, and I particularly enjoyed Alex Kasman's Mathematical Fiction site, which discusses the occurrence of mathematics in fiction: http://math. cofc.edu/faculty/kasman/MATHFICT/default/html. The Web-based encyclopedias Wikipedia (http://en.wikipedia.org) and Eric W. Weisstein's MathWorld, a Wolfram Web Resource (http://mathworld. wolfram.com) are always excellent sources of mathematical information.

Other interesting Web sites, technical and artistic sources, and recommended reading are listed in the references section.

The chapter patent-diagram frontispieces are from *U.S. Pat. 3,648,407* (1972, Introduction), *U.S. Pat. 3,991,631* (1976, Chapter 1), *U.S. Pat. 4,919,427* (1990, Chapter 2), *U.S. Pat. 4,384,717* (1983, Chapter 3), *U.S. Pat. 4,640,029* (1987, Chapter 4), *U.S. Pat. 3,758,981* (1973, Chapter 5), *U.S. Pat. 4,253,836* (1981, Chapter 6), *U.S. Pat. 5,411,330* (1995, Chapter 7), *U.S. Pat. 3,953,679* (1976, Chapter 8), *U.S. Pat. 6,779,936* (2004, Solutions), and *U.S. Pat. 396,658* (1998, References).

MÖBIUS LIMERICKS
TO GET YOU IN THE MOOD*

A young man named Möbius (quite clever),
A circle of paper would sever.
He'd then tie a knot
As part of his plot
To stay in Las Vegas forever.
—*Paul Cleverley*

Quoth mother of four year old Pete:
"You may not cross Möbius Street."
But an easy walk,
Once around the block,
Allowed him to manage the feat.
—*Chuck Gaydos*

There once was a fellow from Trent
Who conversed with a Möbius bent.
On and on he would blather
On this and that matter
With twisting one-sided intent.
—*Quinn Tyler Jackson*

Said the ant to its friends: I declare!
This is a most vexing affair.
We've been 'round and 'round
But all that we've found
Is the other side just isn't there!
—*Cameron Brown*

---

* These limericks are the winners of the Möbius Limerick Contest, which I sponsored while writing this book.

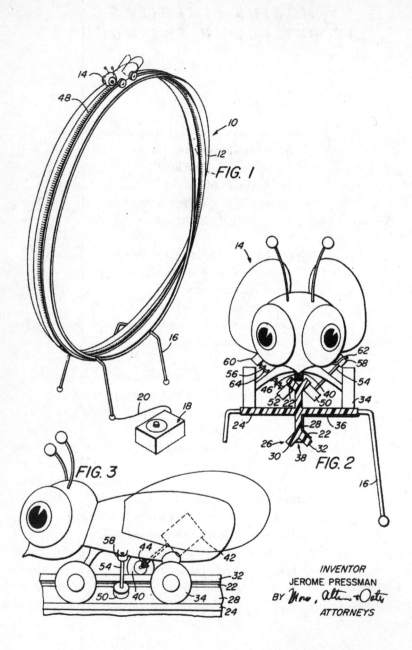

FIG. 1

FIG. 2

FIG. 3

INVENTOR
JEROME PRESSMAN
BY
ATTORNEYS

INTRODUCTION

August Ferdinand Möbius was born on 17 November 1790 and died on 26 September 1868. During the course of his lifetime, the pursuit of mathematics in Germany was transformed. In 1790, it would be hard to find one German mathematician of international stature; by the time he died, Germany was the home and training ground of the world's leading mathematicians . . .

—*John Fauvel, "A Saxon Mathematician,"*
in Möbius and His Band

A Hole through a Hole in a Hole

The universe cannot be read until we have learnt the language and become familiar with the characters in which it is written. It is written in mathematical language, and the letters are triangles, circles and other geometrical figures, without which means it is humanly impossible to comprehend a single word.

—*Galileo Galilei,* Opere, Il saggiatore, *1633*

When I talk to students about topology, the science of geometrical shapes and their relationships to one another, I stretch their minds by sketching several simple shapes. Some look like doughnuts, others like pretzels, and a few like twisted bottles with long necks. I then pose a question to my audience: *Can you imagine a hole through a hole?*

The most common answer is that this is impossible. I smile and respond, "Well, I am about to show you something even better than a hole in a hole. I will show you a hole in a hole in a hole!" With a flourish, I make a sketch of the object in figure I.1, and the audience invariably smiles with delight. Throughout this scrapbook of curiosities, I hope to surprise you with other geometrical treats.

I.1

A playful dog loses his bone in a hole through a hole in a hole.
(Drawing by April Pedersen.)

Topology is about spatial relationships and glistening shapes that span dimensions. It's the Silly Putty of mathematics. Sometimes, topology is called "rubber-sheet geometry" because topologists study the properties of shapes that don't change when an object is stretched or distorted. The best way for people of all ages to fall in love with topology is through the contemplation of the Möbius strip—a simple loop with a half twist (figure I.2)

I.2

A Möbius strip.

The Skull of Dr. Möbius

Even a great mathematician is almost always unknown to the public. His "adventures" are usually so confined to the interior of his skull that only another mathematician cares to read about them.
—Martin Gardner, "The Adventures of Stanislaw Ulam," 1976

In this book I'll frequently digress into topics related to the Möbius strip and topology that you won't find in most math books. For example, just a few months before writing this introduction, I had the opportunity to see the skull of my hero, the mathematician August Ferdinand Möbius, who described the strip that bears his name. The upper half of Möbius's skull appears in a weird 1905 photo published in his grandson's book *Ausgewählte werke* (*Selected Works*) (figure I.3).

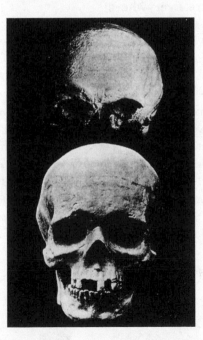

I.3
The skulls of August Ferdinand Möbius (above) and Ludwig van Beethoven (below), from a book by grandson Paul Möbius. Paul disinterred his dad to create this bizarre photo. [Source: Paul Möbius, *Augewähte Werke, Vol. 7, Tafel III*, The British Library, 1905, as displayed in *Möbius and His Band*, p.17, edited by John Fauvel, Raymond Flood, and Robin Wilson. (New York: Oxford University Press, 1993).]

Grandson Paul Möbius, an often-brilliant neurologist, did have a few odd notions, including such archaic ideas that a left fronto-orbital bump, which anatomist Dr. Franz Gall designated the "mathematical organ," was particularly large on August Möbius's head. Today, of course, we don't take Dr. Gall's phrenological ideas seriously. Looking at the photo of August Möbius, I can't tell if the supposed bump existed on his head, but I do know that Paul made a vast study of mathematicians' heads, collecting skull data from men living and dead and including photos in his thorough monograph on this subject. His mission was to demonstrate that mathematical ability was intimately linked to bumps on the head.

Thinking about all these skulls gives me a shiver. Exhumations at the Leibzig cemetery gave Paul the perfect opportunity to dig up his grandfather's skeleton so he could handle the skull and make his observations.

Möbius Strips Are Everywhere!

A mathematician, like a painter or poet, is a maker of patterns. If his patterns are more permanent than theirs, it is because they are made with ideas.
—*G. H. Hardy,* A Mathematician's Apology, *1941*

The Möbius strip has fascinated both mathematicians and laypeople ever since Möbius discovered it in the nineteenth century and presented it as an object of mathematical interest. As the years passed, the popularity and application of the strip grew, and today it is an integral part of mathematics, magic, science, art, engineering, literature, and music. It has become a metaphor for change, strangeness, looping, and rejuvenation. In fact, today the Möbius band is the ubiquitous symbol for recycling, where it represents the process of transforming waste materials into useful resources (figure I.4).

I.4
The ubiquitous symbol for recycling.

The recycling symbol consists of three twisted chasing arrows in the shape of a triangle with rounded vertices. If the correspondence of the symbol to the Möbius band is not clear to you, the similarity will become evident as you read further. What would Möbius have thought if he could look into the future and see that the most common use of his loop was in the area of waste disposal! The recycling symbol was designed in 1970 by Gary Anderson, a student at the University of Southern California at Los Angeles. Anderson submitted his logo to a nationwide contest sponsored by the Container Corporation of America.

Today, the Möbius strip is everywhere—such a compelling shape! Variously called the "Möbius strip" (38,000 Web sites), "Möbius band" (seven thousand Web sites), or "Möbius loop" (11,000 Web sites), interest in the wonderful object is growing. Of course, one cannot take these Googled Web site numbers too seriously, because the phrase may sometimes refer to the name of a rock group or a non-Möbius object.

In this book I will touch on the Möbius strip's appearance in a variety of settings, from molecules and metal sculptures to postage stamps, literature, architectural structures, and models of our entire universe. The strip is featured in countless technology patents, which decorate the frontispieces of each chapter and are briefly covered in chapter 4.

Today, the Möbius strip has become common in jewelry, including popular golden pendants inscribed with Hebrew verses from the Bible. It's the logo for *Möbius: The Journal of Social Change*. It's the name of a Santa Cruz, California, company that specializes in the conservation and restoration of oil paintings. In 2004, Möbius beer, infused with taurine, ginseng, caffeine, and thiamine, went on sale in Charleston, South Carolina, each can emblazoned with the Möbius strip. "Möbius beer will keep you going on and on all night long," says the company literature. Even the calcium dietary supplement Caltrate features a big purple Möbius strip on its packaging.

MÖBIUS is also the name of a poetry magazine, whose logo is a Möbius strip. The "Möbius Flip" is the name of an acrobatic stunt performed by freestyle skiers that involves a twist while somersaulting through the air. The Colorado Ski Museum sells a half-hour videotape titled *The Möbius Flip*, featuring spectacular glacier skiing. In addition, various waterskiing sports feature "Möbius tricks" and related inverted spins on hydrofoil water skis.

Numerous Möbius objects have entered my own personal hall of fame. For example, my favorite wood engraving that features a Möbius band is Dutch artist M. C. Escher's *Möbius Strip II*, which presents red ants crawling on the surface of a Möbius strip. My favorite Möbius strip sculpture is Swiss-born artist Max Bill's *Endless Ribbon*, made of granite and displayed in sculpture gardens in the early 1950s. My favorite movies featuring the strip are *Möbius*, directed by Gustavo Mosquera, and *Eternal Sunshine of the Spotless Mind*, directed by Michel Gondry. We'll discuss Möbius plots in literature and movies in chapter 8 and in this book's conclusion.

These days the Möbius strip has also become an icon for endlessness, and we'll touch on many popular, offbeat, and imperfect Möbius metaphors, as well as geometrical objects that are more precisely identified as Möbius strips. In literature and mythology, the Möbius metaphor is used when a protagonist returns to a time or place with an alternative viewpoint, because a true Möbius strip has the intriguing property of reversing objects that travel within its surface. This geometrical reversal will be made clear in chapter 6.

Perhaps the most common contemporary use of the term "Möbius strip" occurs when alluding to *any* kind of mysterious looping behavior, or as author John Fauvel says, "The cultural pervasiveness of the notion of the Möbius band is now assured because, rather like some other popular mathematical metaphor, it has begun to be used in all kinds of contexts for which it is thoroughly inappropriate." Some of the quotes at the ends of each chapter are examples of these amusing contemporary uses.

Smorgasbord

Geometry is unique and eternal, and it shines in the mind of God. The share of it which has been granted to man is one of the reasons why He is the image of God.

–Johannes Kepler, "Conversation with the Sidereal Messenger," 1610

As in all my previous books, you are encouraged to pick and choose from the smorgasbord of topics. Sometimes, I repeat a definition so that it is easier for you to browse chapters that most interest you. Many of the chapters are brief to give you just the tasty flavor of a topic. Those of you interested in pursuing specific topics can find additional information in the referenced publications. In order to encourage your involvement, the book contains several puzzles (denoted by an ☞ symbol) for you to ponder, with solutions at the end of the book. Spread the spirit of this book by posing these questions to your friends and colleagues the next time you plunk down on the couch to listen to the Möbius Band, a contemporary western Massachusetts music trio playing at the edges of rock, electronic, and experimental music.

Whatever you believe about the possibility of some of the weird shapes in this book and the strange models for the cosmos, my topological analogies raise questions about the way we see the world and will therefore shape the way you think about the universe. For example, you will become more conscious about what it means to visualize a one-sided object in your mind or what it means to have orientation-reversing paths in space.

By the time you've finished this book, you will be able to do the following:

- understand arcane concepts such as paradromic rings and ekpyrotic models of the universe's creation
- impress your friends with such terms as Schulpforta, homeomorphisms, sphere eversions, nonorientable surfaces, Boy surfaces, cross-caps, Roman surfaces, real projective planes, the Möbius function $\mu(n)$, squarefree numbers, Merten's conjecture, the ubiquity of $\pi^2/6$, hexaflexagons, Möbius shorts, Möbius tetrahedra, solenoids, Alexander's horned spheres, prismatic doughnuts, the barycentric calculus, and Bonan-Jeener's Klein bottles
- write better science fiction stories involving the Möbius strip
- understand most people's rather limited view of space and shape.

You might even want to go out and see Eugene Ionesco's play *The Bald Soprano* with its Möbius-like twist, read my novel *The Lobotomy Club*, which centers on a mythical arrangement of brain cells called a cerebral Möbius strip, or buy one of the latest glass Klein bottles available on the Web from Acme Klein Bottle.

Geometry and the Imagination

I could be bounded in a nutshell and count myself a king of infinite space.

—William Shakespeare, Hamlet, 1603

When I receive e-mail from teachers and laypeople about mathematics, I find that the mathematical objects that excite them most are geometrical shapes with startling properties. They are also fascinated by the idea that our universe could comprise a space shaped like a doughnut or include higher dimensions. All students seem to be delighted by the miraculous four-dimensional Klein bottle or by contemplating what it would be like to live on a Möbius band.

Alas, most high school students are never exposed to topology. Hopefully, this book on August Ferdinand Möbius and his band may serve as a brief teaser to more advanced concepts, especially for readers who would never go beyond trigonometry at school or even in technical jobs. Interestingly, although topology grew out of puzzles involving simple objects such as the Möbius band, today the modern topologist wades in a morass of mathematical theory. In fact, some topologist friends are suspicious of theorems that must be visualized to be understood. Martin Gardner notes the following in *Hexaflexagons and Other Mathematical Diversions*.

> People who have a casual interest in mathematics may get the idea that a topologist is a mathematical playboy who spends his time making Möbius bands and other diverting topological models. If they were to open any recent textbook in topology, they would be surprised. They would find page after page of symbols, seldom relieved by a picture or diagram.

In this book, I hope to give readers a taste of topology, higher dimensions, and bizarre twisted forms using very few formulas. Topology is an infinite fountain of strange and wondrous forms, and I've been in love with recreational topology for many years for its educational value. Contemplating the simplest of problems stretches the imagination. More generally, the *usefulness* of mathematics allows us to build spaceships and investigate the geometry of our universe. Numbers and geometry will be our first means of communication with intelligent alien races. It's even possible that an understanding of topology and higher dimensions may someday allow us to escape our universe when it ends in either great heat or cold, and then we could call all of spacetime our home.

Today, mathematics has permeated every field of scientific endeavor and plays an invaluable role in biology, physics, chemistry, economics, sociology, and engineering. Mathematics can be used to help explain the colors of a sunset or the architecture of our brains. Mathematics helps us build supersonic aircraft and roller coasters, simulate the flow of Earth's natural resources, explore subatomic quantum realities, and image far-away galaxies. Mathematics has changed the way we look at the cosmos.

Quotations

A mathematician is a machine for turning coffee into theorems.
—Paul Erdös, quoted in Paul Hoffman's The Man Who Loved Only Numbers

I'm a voracious reader and keep a scrapbook of intriguing quotations that come across my line of sight each day. Many come from newspapers, magazines, and books that I'm reading. At the end of each chapter of this book are snippets from these sources that feature a Möbius strip metaphor in an interesting way. I denote these timely and sometimes quirky quotes with a 🌑 symbol. I welcome your feedback and look forward to your own Möbius quotation submissions. Enjoy!

MÖBIUS STRIP IN RELIGION

🌑 *But God has no skin and no shape because there isn't any outside to him. With a sufficiently intelligent child, I illustrate this with a Möbius strip.*
—Alan Watts, The Book: On the Taboo Against Knowing Who You Are

🌑 *Like the Möbius strip, the inside and outside of God are the same.*
—Frank Fiore, To Christopher: From a Father to His Son

🌑 *Only a Jew can understand that God's will and our free will work hand in hand. It would drive other people crazy. It's like a Möbius strip: it's in and out and up and down, together.*
—Robert Eisenberg, Boychiks in the Hood: Travels in the Hasidic Underground

---

MÖBIUS MAGICIANS

Möbius is a household name—at least, it is in mathematical houses—thanks to a topological toy. But August Möbius influenced mathematics on many levels . . . [His modern legacy] is a large part of today's mathematical mainstream.

—Ian Stewart, "Möbius's Modern Legacy,"
in Möbius and His Band

When I was in third grade, I went to a neighbor's birthday party that featured a magic show. A magician in a tall black hat handed me a band, which it seemed he had made by pasting together the ends of shiny strips to form a long loop of ribbon. He had three such loops—one strip was red, another blue, the third purple. The magician's name was Mr. Magic. Very original.

Mr. Magic smiled as he drew a black line along the middle of each of the long strips, like a dashed line painted on a highway (figure 1.1). He showed the strips to the audience. One kid grabbed, and Mr. Magic said something like, "Patience!"

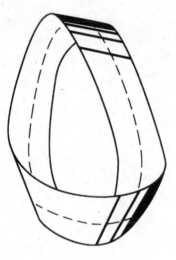

1.1
A Möbius strip with line drawn along the middle.

I was a shy child and well-behaved. Mr. Magic must have sensed that and handed me a scissors. "Young man, cut the strip lengthwise along the line." He motioned along the dashed line on one of the strips.

I was excited and continued to cut the red strip until I reached the starting point of my cut. The red band fell apart to form two totally separate rings. "Cool," I said, but really I wasn't too impressed. Still, I wondered about what was happening.

"Now cut the others."

I nodded. After I cut the blue strip, it formed a single band twice as long as the original. Someone clapped. He handed me the remaining purple strip. I cut this one, and it formed two interlocking rings—like two links of chain.

Each color behaved so differently—now that *was* pretty cool! The bands had totally different properties, although they had looked identical to me. A few years later, a friend explained the mysterious trick to me. The red, blue, and purple strips were each created differently when the ends of the ribbons were joined. The loop of red ribbon was the most straightforward. It was a simple loop with no twist, resembling an ordinary conveyor belt or a thick rubber band. The blue loop, however, was the famous Möbius strip, formed by twisting the two ends of the ribbon 180 degrees with respect to each other before pasting the ends together. This is typically called a "half twist." The purple loop was formed by twisting one end 360 degrees relative to the other before pasting the ends.

Today, magicians often call this stunt the Afghan Bands trick, although I'm not sure where the name originates. The trick, performed under this name, dates back to around 1904.

According to Martin Gardner's *Mathematics, Magic, and Mystery*, the earliest reference for use of the Möbius strip as a parlor trick is the 1882 English edition of Gaston Tissandier's *Les recreations scientifiques*, first published in Paris in 1881. Carl Brema, an American manufacturer of all kinds of magic tricks, frequently performed the Afghan Bands trick in 1920, using red muslin instead of paper. In 1926, James A. Nelson described a method for preparing a paper band so that two cuttings of the band produced a chain of *three* interlocked bands (figure 1.2).

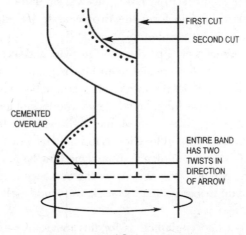

1.2
James A. Nelson's method for preparing a "magical" paper band so that two cuttings of the band produce a chain of three interlocked bands.
(After Martin Gardner, *Mathematics, Magic, and Mystery*.)

Magician Stanley Collins described another fascinating trick in 1948 involving a twisted band and a ring. He placed a small metal ring on a paper or cloth strip and then joined the ends of the strip after three twists to form a closed loop. As usual, the magician cuts along the middle of the band (like cutting along the centerline of a highway) until he reaches the starting point, producing one large strip knotted around the ring.

Today, professional magician Dennis Regling, who performs "gospel magic" for Sunday schools and Bible camps, uses Möbius strip magic for enhancing belief in God. Just like Mr. Magic, Dennis uses the rings in a gospel presentation by calling up three volunteers. Next, he places the large rings over the heads of the volunteers and explains ". . . how God has made us, and though we are alike in many ways, he has given each of us special gifts too. That we are all uniquely special in God's eyes." He cuts the three different loops with scissors, each producing the three outcomes I described previously.

Another professional gospel magician, Eric Reamer, also uses the three loops to promote religion. Eric is part of a national evangelistic ministry designed to bring the "truth of the Gospel of Jesus Christ" to a "needy" world by using visual object lessons and optical illusions. First, he shows his audience the loop with no twists and says, "I love circles! They are so cool! They have no beginning, and no end, and that reminds me of God!" He then describes how Jesus was similarly eternal, and he tears the standard loop to form the two separate but identical loops, which symbolize God the Father and Son.

Next, he presents the loop with the full twist and explains that the Bible teaches us that God created us in his image, and that God "sent Jesus, so that we might ask Him into our hearts, and be eternally together with God!" Eric tears the loop, creating two interlocked loops.

Finally, Eric presents the true Möbius loop with the half twist, and says, "God must have a lot of love for us to send His only Son, don't you think?" He then asks the audience to imagine how large this love must be. He tears the Möbius strip and shows the audience that the loop has doubled in length. Eric says that this trick also lends itself to lessons on fellowship and marriage.

We will delve into explanations for this magic in coming chapters and explore even more unusual shapes, but for now it is amusing to ponder how Möbius's abstract paper in mathematics, which introduced the strip over a century ago, is today used for mystifying children and for gospel magic that attracts children to Jesus and deepens faith in the divine.

◉ Treadmill Puzzle

For this puzzle, let us imagine that Dr. Möbius was a successful but eccentric inventor. During his travels through Saxony, he invents the exercise device shown in figure 1.3. He hopes that he and his heirs will someday make a lot of money with his ingenious machine. But does it really work? As Dr. Möbius runs, will the treadmill turn, or is it locked, thereby causing Dr. Möbius to run off the end and plunge into the deep ravine? What effect does the figure eight belt have on the operation of the device? Would the operation be different if this figure eight were replaced with a Möbius strip (a loop of conveyor belt with a half twist)? If the device does not work, how would you fix it? Would the device function any differently if all belts were twisted? (Turn to the solutions section for an answer.)

1.3
Will the belts on Dr. Möbius's exercise treadmill turn freely if the figure eight belt is replaced by a Möbius strip? (Drawing by Brian Mansfield.)

A Word on Möbius's Place in History

🏵 *It is an accident of history that Möbius's name is remembered because of a topological party-piece. But it was typical that Möbius should notice a simple fact that anyone could have done in the previous two thousand years—and typical that nobody did, apart from the simultaneous and independent discovery by Listing.*

—Ian Stewart, "Möbius's Modern Legacy," in Möbius and His Band

Knots, Civilization, Autism, and the Collapse of Sidedness

A burleycue dancer, a pip,
Named Virginia, could peel in a zip!
  But she read science fiction
  And died of constriction
Attempting a Möbius strip.
                    —*Cyril Kornbluth, "The Unfortunate Topologist,"* 1957

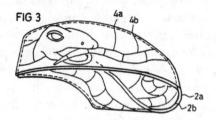

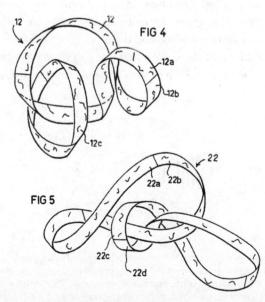

Ants Inside Spheres

If I were to hand you a hollow sphere containing an ant, it's easy to see that the sphere has two distinct sides. An ant walking on the inside of the sphere can't reach the outside surface, and an ant crawling on the outside can't reach the inside.

A plane extending in all directions to infinity also has two sides—an ant crawling on one side cannot reach the other. Even a finite plane, such as a page of paper torn from this book, is considered two-sided if an ant is not allowed to traverse the sharp edges of the boundaries of the paper. Similarly, a hollow doughnut shape, or torus, has two sides. A can of soda has two sides. The first one-sided surface discovered and investigated by humans is the Möbius strip. It seems far-fetched that no one on Earth had described the properties of one-sided surfaces until the mid-1800s, but the history of science and mathematics has recorded no such observations.

A Möbius strip (or band) is a fascinating surface with only one side and one edge. As I suggested in the previous chapter, to create the strip, simply join the two ends of a long strip of paper after giving one end a 180-degree twist with respect to the other end. The result is a one-sided surface—a bug can crawl from any point on such a surface to any other point without ever crossing an edge. In contrast, if you join the ends of the strip without twisting, the result resembles a cylinder or a ring, depending on the width of the strip. Because a cylinder has two sides, you can color one side of the cylinder red and the other green. Try coloring a Möbius strip with a crayon. It's impossible to color one side red and the other green because it has only one side (Figure 2.1). This also means that you can draw a continuous line between any two points on a Möbius strip without crossing an edge.

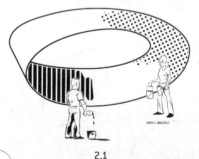

2.1

Attempting to color a Möbius strip. Two painters are confused as they try to paint one side red and one side green. This confusion is actually the key component of a tragicomical story titled "A. Botts and the Möbius Strip," discussed in chapter 8, in which a painter tries repeatedly to paint just one "side" of a Möbius belt.

Construct a Möbius strip yourself right now and place it on a table. Put one finger on one edge and another finger on the "other." Keep one finger stationary as you move the other finger along the edge. Eventually, the moving finger will touch every point along the edge and collide with the stationary one, clearly showing us that the strip has only one edge. In fact, any strip of paper with an *odd* number of half twists is Möbius-like because all such strips have only one surface and one edge.

Dissecting the Band

The Möbius strip has numerous fascinating properties. If you cut along the middle of the strip, as I discussed when referring to the magic tricks in chapter 1, instead of producing two separate strips, you will be left with one long strip with two half twists in it. If you cut this new strip along the middle, you get two strips wound around each other. In other words, this second cutting produces two linked bands.

Alternatively, if you cut along a Möbius strip a third of the way in from the edge, you will get two strips—one is a thinner Möbius strip, and the other is a long strip with two full twists in it (a full twist is a 360-degree twist). Let's try to visualize this. We've learned that if you cut along the *middle* of a Möbius strip, you will return to the starting point of the cut in the middle of the strip. You'll be traversing the strip one time before returning. On the other hand, if you start cutting one third of the way from one edge, you will not meet the start of your cut until you've been around the Möbius strip twice, because on the second time "around" your cut will be a third of the strip's width away from your starting cut along the strip.

In other words, the cutting takes you twice around the Möbius band before you return to your staring point and produces two bands (figure 2.2). Let's call the two resulting strips band *A* and band *B*. Band *A* is identical to the original Möbius band except its width is a third of the original—in fact it is the central third of the original Möbius strip. Band *A* is the smaller of the two bands in figure 2.2. Band *A* is linked with band *B*, which is twice as long as A. Thus, the trisection of a Möbius strip produces the small Möbius band *A* linked to the longer two-sided band *B* that has four half twists.

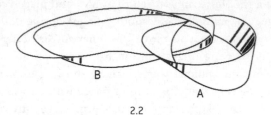

2.2

Cutting along a Möbius strip, a third of the way in from the edge,
produces two strips—one is a small Möbius strip, and
the other is a long strip with four half twists in it.

In *Mathematical Magic Show*, Martin Gardner has shown that it is possible to manipulate the interlocked *A-B* rings from figure 2.2 until they sit together to form a triple-thick Möbius band, as shown in figure 2.3. The darkened edge is the edge of ring *A*.

2.3

2.3 A triple-thick Möbius band can be formed from the A and B rings from figure 2.2.

Let's examine this triple-thick object more closely. In this wonderful nesting, the two outer "strips" appear to be separated all the way around by a Möbius strip sandwiched "between" them. Gardner notes that the same structure can be constructed by putting three identical strips together, holding them as one, giving them a half-twist, and then joining the three pairs of corresponding edges. If you attempt to color the triple-thick band blue on its "outside," you will find it possible to interchange the outside layers so that the blue side of the larger band goes into the interior, and the triple-thick band becomes white on its "outside."

Let's consider other cutting experiments. If you start with a "parent" Möbius band with three half twists and then cut it along the middle to produce a child, you'll generate one larger child band with eight half twists. You can imagine lots of cutting experiments, but we can make a number of generalizations. For example, to calculate the number of half twists in a child, double the number of half twists in the parent band and add two.

Möbius himself considered and drew various variants of the Möbius strip. Figure 2.4 is from Möbius's unpublished writings and shows his band and several twisted relatives. The loop of paper has two sides if the number of half twists is even and one side if the number is odd.

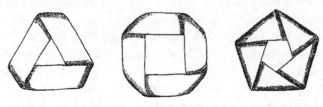

2.4

Möbius's band and several twisted relatives, from Möbius's own unpublished writings. [Source: Möbius's *Werke*, II, page 520. See also page 122 of *Möbius and His Band*, edited by John Fauvel, Raymond Flood, and Robin Wilson. (Oxford University Press, 1993).

We can use mathematical notations to make more generalizations on the cutting properties of twisted strips. Imagine that one end of the paper strip receives m half twists (that is, is twisted through $m\pi$ radians or $m \times 180°$) before it is glued to the other end. If m is even, we create a surface with two sides and two edges. If the strip is cut along a midline between the edges, we obtain two rings, each of which had m half twists and which are linked together $\frac{1}{2}m$ times. If m is odd, we produce a one-sided surface with one edge. If this loop is cut along the midline, we obtain only one ring, but it has $2m + 2$ half twists, and if m is greater than 1, the result is knotted.

Simple Sandwich Möbius Strip

One of the most mystifying Möbius arrangements is the sandwich Möbius strip, created with just *two* strips of paper. I have known people to ponder this for hours while listening to Pink Floyd without ever fully appreciating what they have beheld. Start your construction by placing one strip on top of the other, like two pieces of bread in a sandwich. Together, give the strips both a half twist and tape them as if you were constructing a single Möbius strip (figure 2.5).

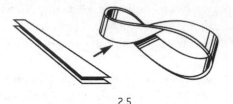

2.5
A sandwich Möbius strip, created from two strips of paper, has remarkable properties.

Hold the double-layer object in your hands. At first, you'll think you have created a pair of nested Möbius strips that hug each other along their shared surface. But how can we truly understand our creation? First, probe the arrangement carefully with a toothpick. Slip the toothpick between the bands. Slowly move it around the bands, and you'll return to your starting point. Yes, it seems perfectly clear that you have two separate bands, for there is always a space between them.

Now, take a red crayon and start coloring one of the Möbius bands. Continue around the entire surface. You will end up returning to your starting point, having twice navigated the sandwich Möbius band, which seems to indicate that the bands are not nested but rather that they are one band with one surface and one edge. For the final shocker, gently pull the two bands apart, and you will discover a single large band with four half twists!

Ljubljana Ribbon, Autism, and Vortex Knots

A Slovenian friend once demonstrated similar kinds of outcomes from cutting exercises presented as a magic trick with a political lesson. In particular, she held up a shining crimson ribbon that, when cut, turned into a trefoil knot, a knot with three crossings (figure 2.6). The trick was supposed to show how individual countries benefited when they came together to form the European Union.

Her crimson ribbon, which she called a Ljubljana ribbon, had three half twists, instead of the usual single half twist for a Möbius strip. When divided lengthwise, the Ljubljana ring turns into the trefoil knot. This conforms to the rule we just discussed: if m is odd, we generate only one ring from cutting the loop, but it has $2m + 2$ half twists, and the result is knotted.

Mathematicians have studied the trefoil knot extensively since the early 1900s. The knot's mirror images are not equivalent, as first proved in 1914 by German mathematician Max Dehn (1878–1952). Dehn wrote one of the first systematic expositions of topology in 1907. (In 1940 he fled Nazi persecution and managed to become the only mathematician ever to teach at Black Mountain College in the U.S.)

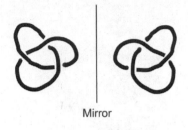

Mirror

2.6

Trefoil knot. The mirror images of this knot are not the same, and no matter how you twist, shift, or deform one of the knots, you cannot make it look like the other unless you cut the knot and retie it.

No matter how you stretch, move, or deform the knots in figure 2.6, you can never transform one into the other. This simple knot got its name from plants of the genus *Trifolium*, which have compound trifoliate leaves. The knot is the basis for countless sculptures and logos, such as the emblem of *Caixa Geral de Depósitos* (the largest bank in Portugal) and John Robinson's trefoil knot sculpture, which resides in the garden of Robinson's studio in Somerset, England (figure 2.7). Note that Robinson's knot is constructed from a ribbon twisted in such a way as to have only one side. The trefoil knot also appears in the famous M. C. Escher engraving *Knots* and in Jos Leys's computer artwork, which are famous for their realistic lighting and shading (figure 2.8)

2.7

Professor Ronnie Brown of the University of Wales, Bangor, and sculptor John Robinson stand before Robinson's trefoil knot sculpture *Immortality*. The work has been adopted by the Department of Mathematics, University of Wales, Bangor, as the departmental logo. (Image courtesy of Edition Limiteé, Geneva.)

2.8
Trefoil knot computer graphics by Jos Leys.

The study of knots, such as the trefoil knot, is part of a vast branch of mathematics dealing with closed twisted loops. For centuries, mathematicians have tried to develop ways to distinguish tangles that *look* like knots from true knots and to distinguish knots from one another. For example, the two configurations in Figure 2.9 represent two knots that for over seventy-five years were thought to represent two distinct knot types. In 1974 a mathematician discovered that it was possible to simply change the point of view of one knot to demonstrate that both knots were the same. Today we call these "Perko pair knots." Although these have been listed as distinct knots in many knot tables since the nineteenth century, New York lawyer and part-time topologist Kenneth Perko showed that they were in fact the same knot by manipulating loops of rope on his living room floor!

2.9
Perko pair knots. Are these knots the same or different?

Two knots are considered to be the same if you can manipulate one of them without cutting so that it looks exactly like the other with respect to the locations of the over- and under-crossings. Knots are classified by, among other characteristics, the arrangement and number of their crossings and certain characteristics of their mirror images. More precisely, knots are classified using a variety of invariants, of which their symmetries is one and their crossing number is another, and characteristics of the mirror image play an indirect role in the classification. No general, practical algorithm exists to determine if a tangled curve is a knot or if two given knots are interlocked. Obviously, simply looking at a knot projected onto a plane—while keeping the under- and over-crossings apparent—is not a good way to tell if a loop is a knot or an unknot. (The unknot is equivalent to a closed loop like a simple circle that has no crossings.) For example, consider the "mystery unknot" in figure 2.10. Can you tell that this is an unknot by manipulating the object in your mind? I asked dozens of colleagues, and most were unable to determine if this was a knot or an unknot simply by looking at it. Could an autistic savant or someone with Asperger's syndrome (high functioning autism) see the solution in his or her mind? Children with autism are sometimes fascinated with items that are not typical toys, such as pieces of string, complex balls of yarn, or rubber bands. Some continually tie knots in strings.

Of the people I surveyed, one woman who could tell this was an unknot, simply by looking, had knitting experience. A woman with Asperger's syndrome was also able to solve this in thirty seconds. She described the process to me as unlooping the string in her head—unwinding it until it became a circle.

2.10
The "mystery unknot." Is this figure a knot?

In 1961 Wolfgang Haken, now of the University of Illinois at Urbana-Champaign, devised an algorithm to tell if a knot projection on a plane (while preserving the under- and over-crossings) is actually an unknot. However, the procedure is so complicated that it has never been implemented. The paper describing the algorithm in the journal *Acta Mathematica* is 130 pages long.

The trefoil knot and the figure eight knot are the two simplest knots, the first having a representation with three crossings and the second with four (figure 2.11). No other knot classes can be drawn with so few crossings. Over the years, mathematicians have created seemingly endless tables of distinct knots. So far, over 1.7 million nonequivalent knots containing sixteen or fewer crossings have been identified.

2.11
Trefoil (left) and figure eight knot (right).

Simple knots like the trefoil and figure eight knots also happen to be the basis for early attempts at a "string theory" for atoms, an area of research that some readers might be surprised to find took place in the nineteenth century. Mathematician and physicist Lord William Thomson Kelvin (1824–1907) accelerated the mathematical theory of knots during his attempts to model atoms, which he suggested were actually different knots tied in the ether that he believed permeated space. He proposed that atoms were actually tiny knotted strings, and the type of knot determined the type of atom (figure 2.12). Physicists and mathematicians of his day set to work making a table of distinct knots, believing they were constructing a table of the elements. Kelvin's definition of a knot was the same as that used by topologists: a knot is a closed curve that does not intersect itself and that cannot be untangled to produce a simple loop. The topological stability and the variety of knots were thought to account for the stability of matter and the variety of chemical elements.

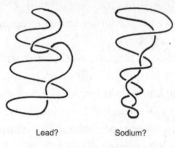

Lead?   Sodium?

2.12

Near the end of the nineteenth century, some scientists believed that each
atom corresponded to a different knot tied in the ether.

Scientists took Kelvin's theory of "vortex atoms" seriously for about
two decades. Even the famous physicist James Clerk Maxwell (1831–1879)
thought that "it satisfies more of the conditions than any [model of the]
atom hitherto considers." Kelvin's vortex atom theory inspired Scottish
physicist Peter Tait (1831–1901) to begin an extensive study and catalogue
of knots to help him understand when two knots were actually different.
However, much of this excitement with knot theory suddenly came to a
halt once scientists discovered that the invisible ether of space did not
exist. Alas, interest in knots continued to wane for decades.

Chemistry has come a long way since the days of Kelvin. Today,
chemists are able to perform the difficult task of actually synthesizing
knotted *molecules*, including molecules with trefoil knots. I'll show you
some of these in chapter 4.

Scientists have also made DNA trefoil and figure eight knots. Closed
circular DNA molecules, such as plasmids, can be knotted, and different
DNA knots can be separated experimentally by a laboratory technique
called gel electrophoresis, in which an electrical current forces molecules
across a span of gel. A molecule's properties determine how rapidly an
electric field can move the molecule through a gelatinous medium.
Knots with different crossing numbers have different speeds of
movement in the gel, and hence produce distinct gel bands.

Entire conferences are devoted to knots today. Scientists study
knots in fields such as molecular genetics—to help us understand how
to unravel a loop of DNA—and particle physics in an attempt to represent the fundamental nature of elementary particles. For example,
Phoebe Hoidn and Andrzej Stasiak of the University of Lausanne,
Switzerland, and Robert Kusner of the University of Massachusetts at

Amherst, study the mathematical complexities of certain knots to develop new theories with the potential to explain properties of elementary particles like electrons.

To understand Hoidn and Stasiak's work, we must first recognize that if a long, practically weightless, silk fiber loop is charged electrostatically (for example, by rubbing it) and then released so that it can relax (ideally, in a gravity-free environment), the ring will form a perfect circle because this balanced shape is a minimum energy configuration. Surprisingly, an electrostatically charged trefoil knot does not form a shape that keeps its three loops as large as possible. Instead, the trefoil knot tightens into a very small region on a perfect circle. This tightening behavior takes place for other kinds of knots as well. In their efforts to think of ways to prevent such tightening, mathematicians are developing models that may one day help us understand the properties of electrons, which are sometimes modeled as little loops of charge, maybe even knotted loops. Within different knot families, Hoidn and Stasiak have found atomlike characteristics such as the quantization of energy (steplike energy differences corresponding to different knots).

Protein biochemists are also fascinated by knots that may reside in large biomolecules. In 2000, British mathematical biologist William R. Taylor developed an algorithm for detecting knots in protein backbones, the coordinates of which are stored in protein databases. In particular, he scanned more than three thousand different protein structures stored in the Protein Data Bank, a worldwide repository of 3-D biological macromolecular structure data.

Taylor found eight knots in his quest. Most of these knots were simple trefoil knots. Several knots were detected in proteins not previously recognized as knotted. One knot occurred in the enzyme acetohydroxy acid isomeroreductase, which was interesting because it sat very deeply in the folded protein, far from the ends of the protein backbone, and in the form of the more complicated figure eight knot. Taylor's 2000 paper in *Nature* describes a protein-folding pathway that may explain how such strange knots are formed. In order to find protein knots, Taylor begins by computationally "holding" the two ends of the protein backbone fixed while the rest of the molecule shrinks until it sometimes forms an obvious knot.

Knots such as the trefoil and figure eight have inspired humans for centuries. A pointy form of the trefoil knot, called a triquetra, was used by the Celtic Christian Church to symbolize the trinity, but the symbol predates Christianity as a Celtic symbol of the triple goddess (Maiden, Mother, and Crone). It's also the symbol of the occult TV show *Charmed*, where it's frequently seen as an ornament hanging from a black cat's collar and represents the three beautiful Halliwell sisters working together as a single force. Back in the 1970s, the triquetra was made famous by its appearance on the jacket of Led Zeppelin's fourth album.

The quintessence of ornamental knots is exemplified by the Book of Kells, an ornately illustrated gospel bible, produced by Celtic monks in about AD 800. It is one of the most lavishly illuminated manuscripts to survive the medieval period. Scattered through the text are letters, animals, and humans, often twisted and tied into complicated knots (figure 2.13). Tightly interlaced bands, knots, and spirals of extraordinary intricacy are everywhere. Computer artist Jos Leys has been inspired by Celtic designs to experiment with various computer renditions, such as the intricate object in figure 2.14. Leys's knot-generation method uses tiles, upon which a simple arrangement of "tubes" is placed. The tiles are then arranged on a grid, like squares on a checkerboard, to form a mosaic containing an intricate knot. Finally, the tile lines are removed to highlight the knotted form. In chapter 7, I will display some even more complicated knots created by mathematicians who experiment on the edges of mathematics and art.

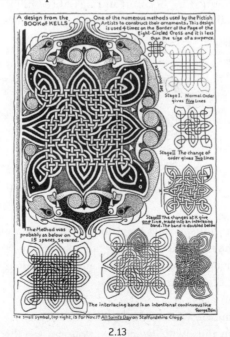

2.13

A design from the Book of Kells from George Bain's *Celtic Art: The Methods of Construction* (New York: Dover, 1971).

2.14
Computer graphics rendition of a complex knot inspired by Celtic designs.
(Created by Jos Leys.)

Another favorite set of interlocking objects of interest to mathematicians and chemists is formed by Borromean rings—three mutually interlocked rings named after the Italian Renaissance family who used them on their coat of arms. Ballantine Beer also uses this configuration in their logo (figure 2.15).

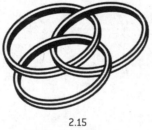

2.15
Borromean rings.

Notice that Borromean rings have no two rings that are linked, so if we cut any one of the rings, all three rings fall apart. Some historians speculate that the ancient ring configurations once represented the three families of Visconti, Sforza, and Borromeo, who formed a tenuous union through intermarriages.

Mathematicians now know that we cannot actually construct a true set of Borromean rings with *flat* circles; you can see this for yourself if you try to create the interlocked rings out of wire, which must be

deformed or kinked to make the shape. The theorem stating that Borromean rings are impossible to construct with flat circles was proved by
Michael Freedman and Richard Skora in their 1987 article "Strange
actions of groups on spheres" [*Journal of Differential Geometry*, 25(1):
75–98]. See also Bernt Lindström and Hans-Olov Zetterström's "Borromean Circles are Impossible," which was published in the 1991 *American Mathematical Monthly* [98(4): 340–341].

In 2004, UCLA chemists created a breathtaking Borromean beauty—
a molecular rendition of interlocked Borromean rings. Each molecule of
the molecular Borromean ring compound was 2.5 nanometers across
and contained an inner chamber that was 0.25 cubic nanometers in
volume and lined by twelve oxygen atoms. The rings include six metal
ions in an insulating organic framework. Researchers are currently
thinking about ways in which they may use molecular Borromean rings
in such diverse fields as spintronics (an emergent technology that
exploits electron spin and charge) or in a biological context such as medical imaging.

Knots and the Triumph of Civilization

It is not an exaggeration to say that knots have been crucial to the development of civilization, where they have been used to tie clothing, to secure
weapons to the body, to create shelters, and to permit the sailing of ships
and world exploration. Knot patterns have been found on burial stones
engraved by Neolithic peoples. The Incas used knots as a form of book-
keeping and as "written language" along strings known as quipu. The
ancient Chinese also used knots for fastening, recording events, and wrapping. The famous Chinese Pan-ch'ang knot, which is actually a series of
continuous loops, symbolizes the Buddhist concept of continuity and the
origin of all things. A few of today's knots have their genesis in the Middle
Ages, when they were used with compound pulleys for lifting and pulling
loads, which were also usually attached with suitable knots.

Sailors used and invented knots to tie ropes to poles, to tie ropes
together, to rig sails, and to hoist loads. Figure 2.16 shows two pages from
a 1943 edition of the U.S. Navy's centuries-old *Bluejacket's Manual* that
features over a thousand pages on such topics as knot tying, signal flags
and pennants, and boat seamanship. In 1902, when Lt. Ridley McLean
first wrote this "sailor's bible," he described it as a manual for every
person in the naval service, from sailor to admiral.

Today, knot theory has infiltrated biology, chemistry, and physics,

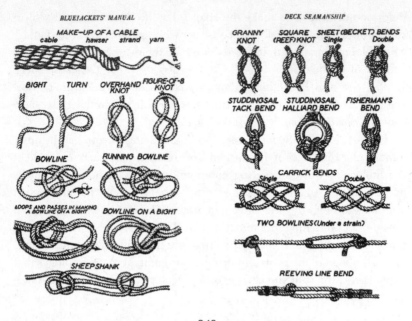

2.16
Two pages from the U.S. Navy's centuries-old *Bluejackets' Manual*.

and in many cases has become so advanced that mere mortals find it challenging to understand its most profound applications. Pick up any modern book on knot theory, and you'll deal with a list of impressive sounding phrases like Conway's polynomial, Conway's skein relation, the HOMFLY polynomial, Jones's polynomials, spin models, Kauffman brackets, finite-order invariants, ambient isotopy, Vassiliev invariants, Gauss diagrams, Knotsevich's theorem, the Yang-Baxter quantum equation, Artin's relation in braid groups, Hecke operator algebra, topological quantum field theory (TQFT), and Temperley-Lieb algebra. In a few millennia, humans have transformed knots from ornamental engravings on rocks to models of the very fabric of reality.

👁 Alien Knot Puzzle

The year is 2050, and Paris Möbius, a descendant of August Möbius, and her girlfriend Nicole are exploring Fifth Avenue in New York City. Suddenly, a race of insectile aliens surrounds them. One of the aliens points at Paris.

"Oh no!" Paris says, her long blond hair shimmering in the sunlight. "What do we do now?"

The tallest of the aliens approaches Paris and points to a loop of rope on the ground (figure 2.17). Then he blindfolds Paris and Nicole and turns to Paris.

"Do you think it is likely that the rope on the ground is knotted?

2.17
A loop of rope. What are the chances that this rope is knotted?

Nicole clenches her fists. "How do we get ourselves into such absurd situations?"

Paris reaches out to hold her hand. "Nicole, don't worry. Even though I glanced at the ground too quickly to notice which segments of rope go over one another, I can determine the exact probability of the rope being knotted. Then I can give the alien an accurate answer."

If you were a gambler, would you bet on the rope being knotted? (Turn to the solutions section for an answer.)

Möbius Strips and Aliens

❦ *We had to bend the Thing into a strange shape to get him through the house doors, a kind of Möbius knot variant. The Thing didn't mind; his body was superfluid anyway . . .*

—*Jeff Noon*, Vurt

❦ *All around her were strange beings. She turned her head and saw that somehow she was in a gigantic room . . . Each tube had what appeared to be a fan belt that was twisted in on itself to form a continuously moving Möbius strip.*

—*Roger Leir*, Casebook: Alien Implants

🖤 *A grin, is that what it was? In an alien, vampire world called Starside on the other side of the Möbius Continuum, there at least it might be called a grin.*

—*Brian Lumley,* Necroscope V: Deadspawn

3 ·

A Brief History of Möbius the Man

*It is paradoxical that modesty and even shyness in Möbius's everyday life
combined in that impressive figure with boldness, fantasy, and abilities.
The mathematical talent of most mathematicians diminishes with age . . .
But time did not diminish Möbius's gifts.*
—Isaak Moiseevich Yaglom, Felix Klein and Sophus Lie

*It is a mathematical fact that the casting of this pebble from my hand alters
the center of gravity of the universe.*
—Thomas Carlyle, *from* Sartor Resartus III

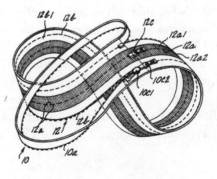

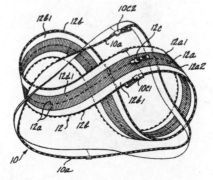

Unfortunately, few detailed English-language biographical accounts on Möbius exist. One excellent introduction to the subject is John Fauvel's "A Saxon Mathematician" in *Möbius and His Band*. Fauvel also points readers to interesting secondary sources on the status of German mathematicians and astronomers during the time of Möbius.

Because Möbius's mother was a descendant of Martin Luther, I was able to reconstruct some of Möbius's family tree by examining lists of approximately 7,900 names, compiled by researchers who record the names and birth dates of Luther's descendants. These old genealogical records allowed me to identify Möbius's children and grandchildren.

August Möbius in a Nutshell

Several members of Möbius's family were both brilliant and famous. In fact, the Möbius family must have had special genes for greatness that became activated with August Ferdinand Möbius (1790–1868), who eventually became a distinguished professor at the University of Leipzig (figure 3.1). Or perhaps the genes came from his wife, Dorothea Christiane Johanna Rothe, who, although completely blind, was able to raise a daughter, Emilie, and two sons, August Theodor and Paul Heinrich (figure 3.2). August Theodor Möbius became one of the world's foremost experts on Icelandic and Scandinavian literature. Grandson Martin August Möbius became professor of botany at the University of Frankfurt and director of Frankfurt's botanical gardens. Great-grandson Hans Paul Werner Möbius was a professor of classical archeology at the Julius-Maximilians University in Würzburg.

3.1
August Ferdinand Möbius (1790–1868). Frontispiece from Möbius's *Werke*.

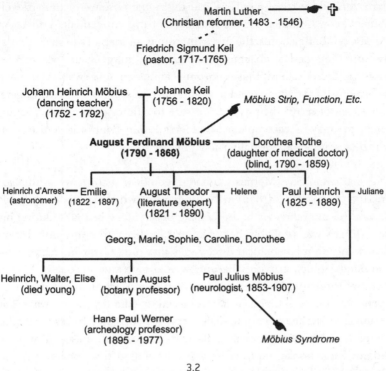

Martin Luther
(Christian reformer, 1483 - 1546)

Friedrich Sigmund Keil
(pastor, 1717-1765)

Johann Heinrich Möbius
(dancing teacher)
(1752 - 1792)

Johanne Keil
(1756 - 1820)

Möbius Strip, Function, Etc.

August Ferdinand Möbius ——— Dorothea Rothe
(1790 - 1868) (daughter of medical doctor)
(blind, 1790 - 1859)

Heinrich d'Arrest —— Emilie   August Theodor —— Helene   Paul Heinrich —— Juliane
(astronomer)   (1822 - 1897) (literature expert)   (1825 - 1889)
                                   (1821 - 1890)

Georg, Marie, Sophie, Caroline, Dorothee

Heinrich, Walter, Elise   Martin August   Paul Julius Möbius
(died young)   (botany professor)   (neurologist, 1853-1907)

Hans Paul Werner
(archeology professor)
(1895 - 1977)   Möbius Syndrome

3.2
August Ferdinand Möbius's family tree.

Möbius's grandson Paul Julius Möbius (1853–1907), whom we discussed in the introduction, became a famous neurologist and psychiatrist. Several of Paul's contributions have been acknowledged by subsequent physicians who gave the Möbius name to various symptoms or illnesses— for example, "Möbius sign," "Möbius syndrome," and "Möbius disease." Despite Paul's genius, he did receive some flack for his pamphlet "The Physiological Mental Weakness of Woman." As a result of its repeated republication, he was accused of disliking women and, as a result, some of his valid contributions to neuroscience may have been looked upon with skepticism.

I first came across the name Paul Möbius several years ago while researching Möbius syndrome, in which children cannot smile. As you can imagine, this facial deficit can cause great hardship in life, and today a few surgeons perform a complex microsurgery operation called the "smile procedure" in order to activate the smile through reattachment of nerves and blood vessels. Möbius syndrome is a rare genetic disorder characterized by facial paralysis, caused by the absence or underdevelopment of two cranial nerves that control eye movements and facial expression. In newborns, the first symptom is an inability to suck. Excessive drooling and strabismus (crossed eyes) may occur. Sometimes, people afflicted with Möbius syndrome cannot smile or swallow, or have deformities of the tongue and jaw, and have missing or webbed fingers. Some cannot move their eyes from side to side or even blink. Möbius syndrome may be accompanied by Pierre Robin syndrome, a disease in which one has an abnormally small jaw.

The mathematician Möbius studied theoretical astronomy with Carl Friedrich Gauss (1777–1855) in Göttingen for two semesters early in his life and became director of the Leipzig observatory in 1848. During his life, Möbius was probably better known for his astronomy popularizations than the mathematical discoveries that today bear his name. The one-sided Möbius strip only became well known after his death.

More generally, Möbius was fascinated by surfaces that could be represented in terms of triangular facets pasted together in various ways. For example, he studied rows of triangles arranged such that the resultant strip could be twisted and joined at its ends to form a one-sided surface. Möbius's notebooks indicated that he developed this concept in September 1858. This discovery of what we now call the Möbius strip was published in an 1865 paper titled "On the Determination of the Volume of a Polyhedron." In that paper, Möbius also proved that polyhedra (multifaceted objects like a tetrahedron) can be imagined that have no volume.

Sometimes, I dream of going back in time to the mid-1800s and visiting Möbius to tell him how famous his band would be one day. The year 1858 was special for many reasons in Europe. Not only was it the year Möbius invented his band, but it was also the year Darwin announced his theory of evolution and Friedrich Nietzsche received a scholarship to an elite preparatory school in Schulpforta, the town of Möbius's birth. And, for a final bit of trivia, in 1858, Hyman L. Lipman of Philadelphia patented a pencil with an attached eraser.

As with many other great works in science and mathematics, Möbius simultaneously discovered the Möbius strip with a contemporary scholar, the German mathematician Johann Benedict Listing (1808–1882). Working independently, Listing first "encountered" the surface in July 1858 and published his findings in 1861. However, Möbius seems to have taken the concept a bit further than Listing by more closely exploring the concept of orientability as it relates to Möbius-like surfaces. We'll discuss orientability in the coming chapters. Möbius also considered numerous other one-sided surfaces, which, as he said, had the "extraordinary" property of giving rise to objects with zero volume. In all of my searches of the literature, I have not been able to find a reference to the one-sided surface prior to Möbius and Listing, which is rather surprising due to the strip's simplicity.

The simultaneous discovery of the Möbius band by Möbius and Listing, just like calculus by Isaac Newton (1642–1727) and German mathematician Gottfried Wilhelm Leibniz (1646–1716), makes me wonder why so many discoveries in science were made at the same time by people working independently. For example, Charles Darwin (1809–1882) and Alfred Wallace (1823–1913) both developed the theory of evolution independently. In fact, in 1858, Darwin announced his theory in a paper presented at the same time as a paper by Wallace, a naturalist who had also developed the theory of natural selection. As another example of simultaneity, mathematicians János Bolyai (1802–1860) and Nikolai Lobachevsky (1792–1856) developed hyperbolic geometry independently and at the same time.

The history of materials science is replete with simultaneous discoveries. For example, in 1886, the electrolytic process for refining aluminum using the mineral cryolite was discovered simultaneously and independently by American Charles Martin Hall (1863–1914) and Frenchman Paul Héroult (1863–1914). Their inexpensive method for isolating pure aluminum from compounds had an enormous effect on industry.

Most likely, such simultaneous discoveries have occurred because the time was ripe for such discoveries given humanity's accumulated knowledge at the time the discoveries were made. On the other hand, mystics have suggested that there is a deeper meaning to such coincidences. Austrian biologist Paul Kammerer (1880–1926) writes, "We thus arrive at the image of a world-mosaic or cosmic kaleidoscope, which, in spite of constant shufflings and rearrangements, also takes care of bringing like and like together." He compared events in our world to the tops of ocean waves that seem isolated and unrelated. According to his controversial theory, we notice the tops of the waves, but beneath the surface there may be some kind of synchronistic mechanism that mysteriously connects events in our world and causes them to cluster.

Möbius: A Mathematical "Watchmaker"

August Ferdinand Möbius was born in 1790 in Schulpforta, Saxony (now Germany). Located in the center of Europe between the cities of Leipzig and Jena (figure 3.3), Schulpforta was a thriving school community where Möbius's father taught dancing. The Saxon town became Prussian in 1815.

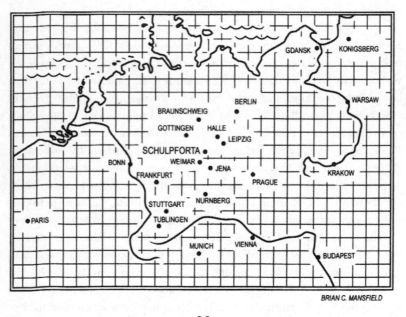

BRIAN C. MANSFIELD

3.3
Schulpforta, the city of Möbius's birth.

Möbius was born during an era of greatness and vast change. Wolfgang Amadeus Mozart (1756–1791) was composing his symphonies in Vienna and died shortly after Möbius's birth. Ludwig van Beethoven (1770–1827) was twenty years old and playing viola. The poet, playwright, and novelist Johann Wolfgang von Goethe (1749–1832) visited Italy several times in the years leading up to 1790, which contributed to his zeal for poetical forms in such plays as *Iphigenie auf Tauris* (1787) and *Torquato Tasso* (1790). And across the Atlantic, Ben Franklin died in the year 1790.

Möbius's mother was a descendant of Martin Luther, the German theologian whose teachings inspired the Protestant Reformation. His father, Johann Heinrich Möbius, died when August was about three years old, and his date of death is given as either 1792 or 1793.

Möbius was fascinated by mathematics at an early age but had no formal education outside the home until he was thirteen years old. By the time he was sixteen, French troops defeated Prussia and Saxony at the Battle of Jena, just a few miles from his home. Although this defeat in 1806 was shocking and demoralizing, it caused a renaissance in German culture. To regain their honor and prestige, the Germans began a period of self-examination and started to reform their economic and educational programs. Schools and teachers became central in society. High school teachers were now held in high esteem. The teaching of mathematics particularly rose in importance and status.

In 1809, Möbius graduated from college, and he became a student at the University of Leipzig, one of the oldest German universities. As with so many families today, Möbius's family also yearned for him to study for a prestigious profession, like law. Möbius acquiesced to his family's demands for his first year of study, but then his passion for mathematics, astronomy, and physics overwhelmed him, and he decided it was better to follow his heart than please his family. He soon became a gifted mathematician and astronomer.

In 1813, Möbius traveled to Göttingen, where he studied astronomy under the world-famous mathematician Carl Friedrich Gauss. According to Calvin Clawson, in his book *Mathematical Mysteries*, Gauss considered Möbius his most talented student.

Many of the great mathematicians of Möbius's time were astronomers–astronomy being a more highly respected and scientific profession before pure mathematics blossomed. Möbius's doctoral thesis, *The Occultation of Fixed Stars*, was followed by a postdoctoral thesis, *Trigonometrical Equations*. (Occultations occur when a moving object, such as a planet or the moon, blocks the light coming from a more distant object, such as another planet or star.) Around this time, he was almost drafted into the Prussian army. He resisted, writing that the idea of the draft, and the draft of *him* in particular, was the "most horrible idea" he had ever heard, and he threatened that anyone who would "venture, dare, hazard, or have the audacity to propose it" would not be safe from his swift and sharp dagger. He never did have to serve in the army.

Möbius pursued his passions and was appointed to the chair of astronomy and higher mechanics at the University of Leipzig in 1816. Alas, he was not as brilliant a lecturer as he was a mathematician, a deficit that slowed his promotion to full professor. In 1820 he married and later had three children.

According to Isaak Moiseevich Yaglom, author of *Felix Klein and Sopus Lie*, Möbius's entire life passed essentially in one city and one building. His study in Göttingen and two or three short excursions through Germany in his youth were his principal "adventures." For the most part, despite his anti-militaristic vitriol, Möbius was quiet, thoughtful, and reserved—a loner who carefully worked through his mathematical arguments. His attention to detail is evident in his various mnemonics that helped him with his schedule and memory. For example, before going out for a walk, he recited the German formula "3S *und Gut*," which helped him remember the first letters of objects that he wanted to take: *Schlüssel* (key), *Schirm* (umbrella), *Sacktuch* (handkerchief), *Geld* (money), *Uhr* (watch), and *Taschenbuch* (notebook). His life was the epitome of regularity. Each night, he wrote in a scientific diary, and through this diary we can trace the evolution of his thoughts.

Möbius's work habits, personality, and personal life greatly affected his mathematical career. For instance, his poor teaching skills repelled potential fee-paying students from taking his classes. Thus, he had to offer his courses for free in order to acquire some students. This is reminiscent of other great minds who may have been poor teachers. For example, so few students went to hear Isaac Newton's lectures at Cambridge that he often read to the walls. Finally, in 1844 the University of Leipzig offered Möbius a professorship in astronomy.

Möbius was a homebody. His life centered exclusively on his studies and his family. This focus on family may have shaped his academic life in that few people read his papers, even though his work was original. Möbius sometimes found that others discovered the same ideas as him years later, publishing them totally unaware of his work.

Finally, Möbius was a numerical watchmaker, working slowly and methodically. Each of his mathematical ideas functioned like a gear that must mesh with other gears with utmost precision. According to biographer Richard Baltzer, editor of *August Möbius, Gesammelte Werke*,

The inspirations for his research he found mostly in the rich well of his own original mind. His intuition, the problems he set himself, and the solutions that he found, all exhibit something extraordinarily ingenious . . . He worked without hurrying . . . almost locked away until everything had been put into its proper place. Only after such a wait did he publish his perfected works . . .

Möbius's Research

Möbius worked from 1827 to 1831 on analytical geometry, projective transformation, and mathematical edifices that are now famous and referred to as "Möbius nets," the "Möbius function," "Möbius statics," the "Möbius transformation," and "Möbius inversion formulas." His paper *Uber eine besondere Art von Umkehrung der Reihen* introduced the endlessly fascinating Möbius function involving just the numbers -1, 0, and +1. We will discuss this function in chapter 5 in great detail.

Möbius was also interested in the mathematics of map coloring, as evidenced by the little problem he posed in 1840. It goes like this: Years ago there was a king with five sons. In his will, the king stated that on his death his kingdom should be divided by his sons into five regions in such a way that each region should have a common boundary with the other four. Can the terms of the will be satisfied? The answer is no. Some popular math books say that Möbius posed, for the first time, the four-color map conjecture stating that four colors are sufficient for the unambiguous construction of any map on a plane. However, the "five sons" problem has less to do with the four-color conjecture than casual inspection. If the answer had been yes, then the four-color conjecture would be false.

We'll discuss the vast variety of Möbius's contributions to mathematics in chapter 5. We've already mentioned a few of his works in astronomy, such as his thesis on the occultations of fixed stars. Here is a sampling of some of the titles of his mathematical papers and treatises:

- 1815, An analytical disquisition on certain peculiar properties of trigonometrical equations
- 1827, The barycentric calculus
- 1829, Metrical relations in the area of linear geometry
- 1829, Proof a new theorem in statics, discovered by Mr. Charles
- 1831, Development of the conditions of equilibrium between forces acting on a free solid body
- 1833, On a special type of dual proportion between figures in space
- 1837, On the midpoint of non-parallel forces
- 1837, Textbook of statics
- 1838, On the composition of infinitely small rotations
- 1840, Application of statics to the theory of geometrical relationships
- 1847, Generalization of Pascal's theorem concerning a hexagon inscribed in a conic section

- 1848, On the form of spherical curves which have no special points
- 1849, On the law of symmetry of crystals and the application of this law to the division of crystals into classes
- 1850, On a proof of the parallelogram law of forces
- 1851, On symmetrical figures
- 1852, Contribution to the theory of the solution of numerical equations
- 1853, On a new relationship between plane figures
- 1853, On the involution of points in a plane
- 1854, Two purely geometrical proofs of Bodenmiller's theorem
- 1855, The theory of circular transformations in a purely geometrical setting
- 1855, On involutions of higher order
- 1856, Theory of collinear involution of pairs of points in a plane or in space
- 1857, On imaginary circles
- 1858, On conjugate circles
- 1862, Geometrical development of the properties of an infinitely thin bundle of rays
- 1863, The theory of elementary relationships
- 1865, On the determination of the volume of a polyhedron

Works published after Möbius's death: On the theory of polyhedra and elementary relationships, Theory of symmetrical figures, On an acoustical problem, On the calculation of the reserve funds of a life insurance company, and On geometrical addition and multiplication.

The Death of Möbius

During Möbius's life (1790–1868), mathematical practice and prestige in Germany grew tremendously. Early in Möbius's life, Germany had very few mathematicians of distinction, but by the time he died, Germany reigned supreme with respect to mathematics and famous mathematicians. This German mathematical enlightenment was facilitated by Prussia's uniting of numerous independent German states, some of which had had a history of acrimony and war. At its peak, Prussia stretched across the north German plain from the French, Belgian, and Dutch borders on the west to regions near the Lithuanian border and to territories that are now in eastern Poland.

Möbius died after having celebrated fifty years of teaching at Leipzig. His beautiful blind wife Dorothea had died nine years earlier.

After his death, historians of science rediscovered Möbius's memoir presented to the Académie des sciences, in which he discussed the properties of one-sided surfaces such as the Möbius strip, which he had discovered in 1858. Möbius must have had no inkling that his name would be forever immortalized by a little strip of paper with a twist, now used in countless arenas.

August Ferdinand Möbius Timeline

Here are many of the important dates in Möbius's life:

- 1790, Born in Schulpforta, Saxony
- 1793, Father died
- 1809, Student at Leipzig University
- 1813, Traveled to Göttingen. Studied with Gauss.
- 1815, Completed doctoral thesis on occultation of fixed stars
- 1816, Appointed extraordinary professor of astronomy, Leipzig
- 1818–1821, Supervised the Leipzig observatory
- 1820, Married Dorothea Rothe
- 1821, Son August Theodor born
- 1822, Daughter Emilie born
- 1825, Son Paul Heinrich born
- 1827, Member, Berlin Academy of Sciences
- 1827, Published *The Barycentric Calculus*
- 1831, Published paper introducing the Möbius function
- 1834–1836, Wrote popular astronomy treatises
- 1837, Wrote textbook on statics (two volumes)
- 1844, Appointed full professor in astronomy, Leipzig
- 1848, Appointed director of the Leipzig observatory
- 1853, Grandson Paul Julius Möbius born
- 1858, Discovered the Möbius band
- 1859, Wife Dorothea died
- 1868, August Möbius died in Leipzig

The False Dawn Animal

Because this book is a "scrapbook" of ideas that interest me, I conclude this chapter with a digression into the life of another famous Möbius, Karl August Möbius (1825–1908), an eminent German zoologist and marine biologist. Many of August Ferdinand Möbius's descendants pursued careers in medicine or the natural sciences. We have Martin the botany professor, Hans the archeologist, and Paul the neurologist. Karl, the marine biologist, was born the same year as one of August Möbius's children, yet I don't think Karl was a descendant of our mathematician Möbius. Still, Dr. Karl Möbius's leadership role in Germany, like other Möbiuses employed in the sciences, intrigues people like me who are fascinated by the Möbius name, which today denotes learning and brilliance. Karl August was the son of Gottlob Möbius, a wheelwright, and Sophie Kaps, and was responsible for constructing Germany's first public aquarium. He was also one of the world's foremost experts on whale anatomy and the formation of pearls. His biggest claim to fame, however, was his discovery that *Eozoon canadense*, which had been considered a living creature, was actually a mineral aggregate! Today, the mysterious biomorphic aggregate is called "the false dawn animal."

Many people were fooled by the lifelike quality of *Eozoon canadense*. For example, in his 1864 presidential address to the British Association for the Advancement of Science, Sir Charles Lyell singled out this fossil as "one of the greatest geological discoveries" of his time. Charles Darwin, in the fourth edition of *The Origin of Species* (1866), was delighted to be able to cite *Eozoon* as the first fossil evidence that the succession of life on Earth proceeded from simple unicellular organisms to complex multicellular animals and plants.

The *Eozoon* scandal started when Sir John William Dawson, one of the foremost geologists in the mid-1800s, concluded that *Eozoon* was actually the shell of a single-celled protistan, complete with chambers and canals, but hundreds of times larger than any of the living forms of his day. In 1865, he formally named the putative fossils *Eozoon canadense*, the "dawn animal of Canada." Others claimed that *Eozoon* was inorganic and merely a layering of minerals in marble. For a decade, increasingly heated debates raged. Finally, in 1879, Karl Möbius demonstrated that *Eozoon* was not a creature at all—it did not have a single characteristic of a protistan. In a few years, virtually no one believed in the dawn animal, which quietly faded away into the sunset.

The *Eozoon* story is one of many in the history of science that suggests that scientific discovery may be symbolized as a Möbius strip stretching through time. In many ways, knowledge moves in an ever-looping Möbius strip. With each journey around the loop, we gaze at the universe and see previous knowledge from a new perspective as theories mutate and new ones form. Some scientific laws may be widely accepted for centuries but later need revision or caveats. In fact, science progresses mainly because theories and laws are never complete. Newton's Law of

Gravity allows us to predict the motions of bullets and cannon balls. However, it does not accurately predict the bending of light rays that pass the Earth. This observation requires us to invoke Einstein's general relativity, which generalizes Newton's Law. Thus, scientific laws generally define what humans know about the world at a point in time. The laws are crucial in defining the workings of reality. However, scientific laws function like oil paintings in which an artist may initially indicate significant visual themes, but brush strokes are still to be added. If no more brush strokes were to come, scientific progress would end. I think Isaac Asimov had the right idea about the future of knowledge when he wrote in his autobiography *I. Asimov: A Memoir*, "I believe that scientific knowledge has fractal properties; that no matter how much we learn, whatever is left, however small it may seem, is just as infinitely complex as the whole was to start with. That, I think, is the secret of the Universe."

☜ Möbius Maze Puzzle

Of the thousands of mazes I have studied both as a child and adult, my favorite maze is "The Möbius Maze" from the book Mind-Boggling Mazes *by Dave Philips (figure 3.4). Start at one worm and find the other by crawling along the pathways as they pass over and under. You must keep in mind which side of the path you are on, and you may not crawl over an edge. (Turn to the solutions section for an answer.)*

3.4
"The Möbius Maze" from Dave Phillips's *Mind-Boggling Mazes* (New York: Dover, 1979). Visit his Web site at www.ebrainygames.com.

Möbius Strip and Licentiousness

🎙 *The Möbius strip has the advantage of showing the inflection of mind into body and body into mind, the ways in which, through a kind of twisting or inversion, one side becomes another. This model . . . provides a way of problematizing and rethinking the relations between the inside and the outside of the subject, its psychical interior and its corporeal exterior, by showing not their fundamental identity or reducibility but the torsion of the one into the other, the passage, vector, or uncontrollable drift of the inside into the outside and the outside into the inside.*

—Elizabeth Grosz, Volatile Bodies: Toward a Corporeal Feminism

🎙 *The road of excess leads to the palace of wisdom . . . and the road is a Möbius strip.*

—Bana Witt, Möbius Stripper

🎙 *All the controls come off, and out pour looks and words and doings and imaginings and a Möbius strip of two bodies gliding and sliding over each other that is a wonder to behold. The power of it makes it a little embarrassing to speak of afterward as you're eating cookies in the kitchen, thinking, "What the hell was THAT?!"*

—Robert M. Alter and Jane Alter, How Long Till My Soul Gets It Right?: 100 Doorways on the Journey to Happiness

TECHNOLOGY, TOYS, MOLECULES, AND PATENTS

While holding the side AB *fixed, twist the strip through an angle of 180°*
about its middle line parallel to A´B´ *, until* A´B´ *is opposite* AB, *and then*
bring A´B´ *into coincidence with* A.
  —August Möbius, "One-Sided Polyhedra," in Gesammelte Werke

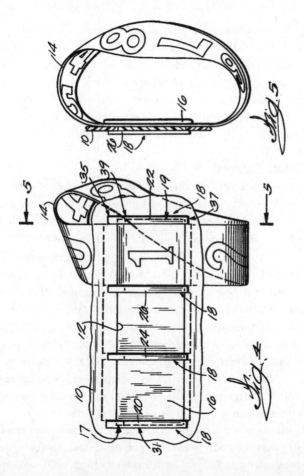

In Roger Zelazny's novel *Doorways in the Sand,* the protagonist encounters a "Rhennius machine," an alien device featuring a Möbius strip. At first, the function of the device is unclear, but it seems to be able to mirror, reverse, or turn inside out objects that pass through its aperture. Humanity received the machine in a technology exchange with aliens.

Beside/below me, where I dangled but a couple of feet above the floor, hummed the Rhennius machine: three jet-black housings set in a line on a circular platform that rotated slowly in a counterclockwise direction, the end units each extruding a shaft—one vertical, one horizontal—about which passed what appeared to be a Möbius strip of belt almost a meter in width.

When I first read about the Rhennius device over a decade ago, I had wondered if the Möbius strip was used in any real devices today. I became obsessed as I searched through the U.S. patent archive and slowly uncovered countless practical applications in modern technology. This chapter, motivated by my interest in Zelazny's fanciful Rhennius device, is a result of my continuing quest to catalogue every major application of the Möbius strip in contemporary devices.

Mathematical Patents

According to a recent article in the *Economist,* the number of patent applications to the U.S. Patent and Trademark Office (PTO) is growing at around 6 percent a year. The wait for a decision is on average twenty-seven months—and much longer for complex applications in advanced sciences. In 2003 the PTO received around 350,000 applications, and in 2004 the PTO had a backlog of over half a million applications to review. Other countries are also encountering a phenomenal growth of submissions. For example, applications to China's patent office increased fivefold from 1991 to 2001.

Mathematical formulas and geometrical shapes can't be patented; however, if an *application* of mathematics and geometry is new, useful, and unobvious, a patent may be obtainable. Also, if a shape has artistic merit and is different from other known shapes, an inventor may obtain a design patent. Design patents have limited value for inventors because variations on a design may not infringe on the patent.

Patents that rely on mathematical shapes are quite common and on the rise. For example, many dozens of patents focus on novel applications of the dodecahedron (an object with twelve pentagonal faces)—from toys to neutron spectrometer monitors for aircraft, to modular buildings, to self-supporting biocompatible porous structures used as a bone mass replacement (U.S. Pat. 6,206,924, issued 2001) (figure 4.1).

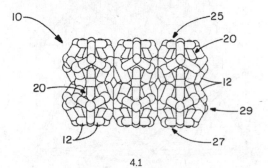

4.1

A figure from the patent "Three-dimensional geometric bio-compatible porous engineered structure for use as a bone mass replacement or fusion augmentation device" (i.e., dodecahedral scaffolding in which bone may grow).

Other patents emphasize the figure eight shape known as the lemniscate in everything from military antennas (U.S. Pat. 6,255,998, issued 2001) (figure 4.2) to baby pacifiers (U.S. Pat. 6,514,275, issued 2003).

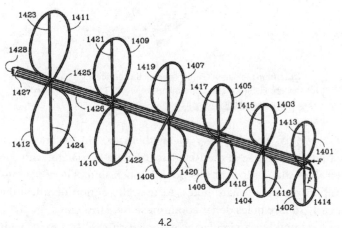

4.2

A figure from the patent "Lemniscate antenna element."

The diamondlike astroid curve is featured in "Cam race for a roller clutch" (U.S. Pat. 4,987,984, issued 1991), and various kinds of polyhedra appear in patents ranging from golf ball dimples (U.S. Pat. 6,749,525, issued 2004) to supports for parabolic reflectors (U.S. Pat. 4,295,709, issued 1981) (figure 4.3).

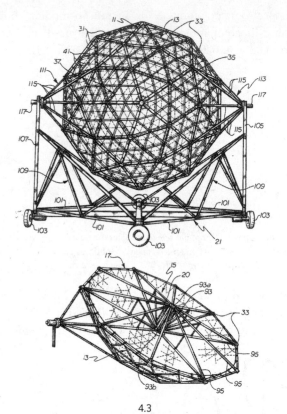

4.3

A figure from the patent "Parabolic reflector comprising a plurality of triangular reflecting members forming a reflecting surface supported by a framework having a particular geometric pattern."

One of my favorite uses of mathematics in patents occurs in numerous applications of the Reuleaux triangle—a triangle with specially curved sides that I show in this book's conclusion. These patents focus on drill bits that cut square holes. At first, the notion of a drill that creates nearly square holes defies common sense. How can a revolving drill bit cut anything but a circular hole? But such drill bits exist, with their cross sections defined by the Reuleaux triangle, named after the distinguished mechanical engineer Franz Reuleaux (1829–1905). For example, figure 4.4 is from the 1978 patent for a "Square hole drill" U.S. Pat. 4,074,778). The Reuleaux triangle also appears in patents for other drill bits, as well as novel bottles, rollers, beverage cans, candles, rotatable shelves, gearboxes, and cabinets.

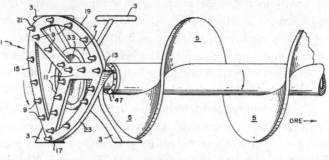

4.4

A figure from the 1978 patent (U.S. Pat. 4,074,778), which describes a drill bit for drilling a square hole based on the Reuleaux triangle.

For the mathematically inclined reader, it is easy to construct a Reuleaux triangle. First, draw an equilateral triangle of side length r. Next, center a compass on each vertex (corner), and draw an arc the short distance between the other two vertices. The perimeter of the Reuleaux triangle will consist of the three connected arcs and resembles the curved triangle at the left of figure 4.4.

Many mathematicians have studied the Reuleaux triangle, so we know a lot about its properties. For example, its area is

$$A = \tfrac{1}{2}\, (\pi\text{-}\sqrt{3})r^2$$

and the area drilled by this kind of drill bit covers 0.9877003907 . . . of the area of an actual square. The small difference occurs because the Reuleaux drill bit produces a square with very slightly rounded corners.

A Smorgasbord of Möbius Strip Patents

The Möbius strip has had countless applications in technology, chemistry, and engineering. Several patents have been granted for Möbius strips used in conveyor belts designed to wear equally on both sides, in toys, and in electronic devices.

One of the earliest patents is Lee De Forest's 1923 U.S. patent for a Möbius filmstrip that records sound on both "sides." A similar concept was later applied to tape recorders so that a twisted tape would run in a continuous loop twice as long as it would otherwise.

Möbius patents started to take off in the late 1940s and early 1950s with the issuing of Owen H. Harris's 1949 patent for a Möbius abrasive belt that greatly increases the polishing or abrading surface (U.S. Pat. 2,479,929) (figure 4.5). Harris describes a Möbius belt with an abrasive surface—typically made with a coating of flint, garnet, corundum, or silicon carbide— that is on both sides of the belt and can be uniformly presented to an object without changing the belt. Harris claims that the belt's increased polishing surface could be used to reduce space in abrading machines by providing a shorter abrasive belt that can do the job of a bigger ordinary belt. Thus, with his invention, Harris succeeded in extending the life of an abrasive belt by presenting a greatly increased abrading surface without lengthening the belt. He wrote, "If desired, a belt of just one-half the length may be used in many installations requiring a specific abrading area since the abrading area is doubled without changing the belt."

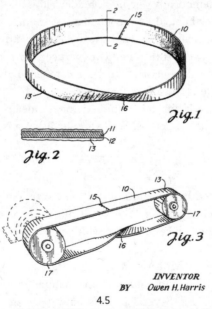

4.5

A figure from Owen H. Harris's 1949 patent on a Möbius abrasive belt that greatly increases the polishing or abrading surface.

Huge Möbius strips have been used as conveyor belts that are more durable because the wear is distributed over the entire surface area. The B. F. Goodrich Company patented a conveyor belt in the form of a Möbius strip that lasts twice as long as conventional belts. In 1957, James O. Trinkle, who worked for B. F. Goodrich, secured a patent for a flexible Möbius conveyor belt used to carry hot materials such as cinders and foundry sand

(U.S. Pat. 2,784,834). In his patent, Trinkle writes that the belt has a longer life when exposed to contact with the heated material. As the belt turns over once during each passage of the belt about pulleys, it alternately presents its opposite heat-resistant face to carry the hot material. Figure 4.6 includes a side view of Trinkle's conveyor belt. The Möbius twist occurs at the location marked 35, aided by guide rollers 33 and 34.

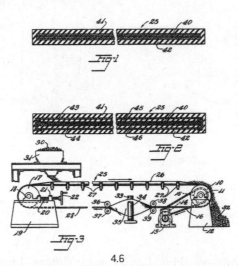

4.6

A figure from James O. Trinkle's 1952 patent for a flexible Möbius conveyor belt used to convey hot materials such as cinders and foundry sand.

The 1960s brings us Möbius patents in more diverse areas, from dry cleaning machines to electrical components. For example, in 1964, Richard Davis invented a Möbius strip nonreactive resistor (U.S. Pat. 3,267,406, issued 1966) (figure 4.7). Davis's employer, who owned the patent, was the United States Atomic Energy Commission, which was established almost a year after World War II ended to control the peace-time development of atomic science and technology. Davis's arrangement resembled a triple-thick Möbius band in which a nonconductive strip is surrounded by metal foil. Davis found that when electrical pulses flowed in both directions around the foil, the strip had interesting electrical properties. He envisioned this to be useful for high voltage, high frequency circuits, especially in pulse applications such as radar, for which the design and operation of these circuits is greatly affected by "unknown reactance in the circuit components themselves or in unwanted coupling between components."

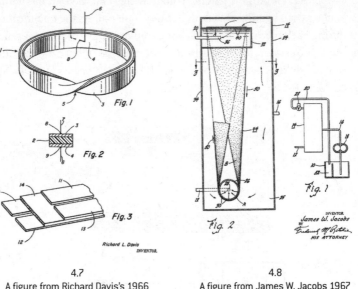

4.7

A figure from Richard Davis's 1966 patent for a Möbius strip electrical resistor.

4.8

A figure from James W. Jacobs 1967 patent for a Möbius self-cleaning filter belt for dry cleaning machines.

In 1967, while working for General Motors, James W. Jacobs patented a Möbius self-cleaning filter belt for dry cleaning machines (U.S. Pat. 3,267,406) (figure 4.8). In dry cleaning machines, the efficiency of operations depends on the effectiveness of filter elements to remove contaminants from the circulating dry cleaning solvent. In the Möbius self-cleaning filter, Jacobs provides a belt loop with a half twist, each section of which is sequentially drenched and drained so that contaminants are first filtered from the liquid onto the Möbius filter and then flushed from the filter element. Jacobs's configuration makes it possible to easily wash dirt and lint from both "sides" of the Möbius filter belt.

In 1986, Thomas Brown secured a patent for a Möbius capacitor (U.S. Pat. 4,599,586) (figure 4.9). Brown uses the Davis Möbius resistor as the interior substructure of his capacitor. In particular, the Möbius capacitor is constructed by layering the continuous conductive surface of the Möbius resistor with a second dielectric material, then layering this dielectric with two separate conductive surfaces in such a way that the conductive surfaces are diametrically opposite each other.

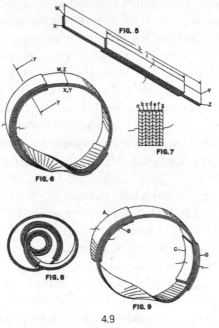

4.9

A figure from Thomas Brown's 1986 patent for a Möbius capacitor.

Today, the Möbius strip is invaluable in numerous toys and puzzles and in many kinds of technological advancements. The Möbius strip plays a role in fun mazes with rolling marbles (U.S. Pat. 6,595,519, issued 2003), in power transmission belts (U.S. Pat. 3,995,506, issued 1976), in small circuit containers that are static resistant (U.S. Pat. 4,766,514, issued 1988)—and it even has the potential for saving lives. As one example, in 2004, John Pulford and Marco Pelosi, working for the Apple Medical Corporation in Massachusetts, patented abdominal surgical retractors with Möbius rings that provide special kinds of torque required to manipulate the retractors during an operation (U.S. Pat. 6,723,044).

Dozens of patents exist with "Möbius" in the title, including inventions for toy puzzles, band saw blades, long-lasting typewriter ribbons, and even a particle accelerating grid. Several patent figures are presented throughout this book to suggest the diversity of inventions. To give you a further idea of the variety of Möbius inventions and the ingenuity of their inventors, the following is a list of several Möbius patents from 1971 to 2004.

- *U.S. Pat. 6,779,936* (2004) "One-sided printing and manufacturing of a Möbius strip" by Ross Martin of Connecticut. The inventor describes methods for the convenient manufacturing of seamless Möbius strips for use in a variety of retail gift items and as a marketing tool for companies promoting "cyclical concepts."
- *U.S. Pat. 6,607,320* (2003) "Möbius combination of reversion and return path in a paper transport system," by Daniel Bobrow et. al. of California and assigned to the Xerox Corporation. Möbius configurations are used in various printing technologies.
- *U.S. Pat. 6,474,604* (2002) "Möbius-like joining structure for fluid dynamic foils" by Jerry Carlow of Texas. Möbius shapes are applied to aircraft and related structures.
- *U.S. Pat. 6,445,264* (2002) "Möbius resonator and filter" by Jeffrey Pond of Virginia and assigned to the U.S. Navy. Möbius configurations are used in electric circuits and, more generally, in the field of electromagnetics.
- *U.S. 6,217,427* (2001) "Möbius strip belt for linear CMP tools" by Christopher Case et. al. of New Jersey and assigned to Agere Systems Inc. The inventors describe a Möbius belt for polishing surfaces. In particular, the invention is used for the "chemical-mechanical polishing of silicon wafer substrates used in fabricating integrated circuits." The belt is constructed as a flexible Möbius strip of a rubberized urethane.
- *U.S. Pat. 5,557,178* (1996) "Circular particle accelerator with Möbius twist" by Richard Talman of New York and assigned to the Cornell Research Foundation, Inc. The inventor describe a circular particle accelerator with a twisted element at one location that gives the accelerator various unique properties. The inventor says that "two traversals of the ring are required to return the particle to a corresponding state, thus the accelerator is termed a 'Möbius' accelerator."
- *U.S. Pat. 5,411,330* (1995) "Möbius shaped mixing accessory" by Yury Arutyunov et. al. of the Russian Federation and assigned to Novecon Technologies. A Möbius-shaped mixing blade is mounted to a shaft. A mirror-image Möbius shaped mixing blade is mounted to a second nearby shaft.
- *U.S. Pat. 5,324,037* (1994) "Möbius strip puzzle" by Ewell Greeson of Georgia. The inventor describes a puzzle game in the shape of a Möbius strip containing several columns and rows. The solution to the puzzle consists of words or phrases spelled by

aligned letters, or a predetermined pattern formed from aligned colors or symbols.

- *U.S. Pat. 4,968,161* (1990) "Ribbon cassette for reinking only one longitudinal half of a Möbius ribbon" by Yoshio Kunitomi et. al. of Tanashi Japan and assigned Citizen Watch Company. An endless inked Möbius ribbon is used for printing on paper.

- *U.S. Pat. 4919427* (1990) "Möbius ring puzzle" by Itzhak Keidar et. al. of Tel Aviv, Israel. This puzzle includes strips of flexible material in the form of twisted loops.

- *U.S. Pat. 4,766,514* (1988) "Pseudo-Möbius static-resistant circuit container" by Kevin Johnson of California. A Möbius-strip container is used to shield electronic circuits.

- *U.S. Pat. 4,640,029* (1987) "Möbius strip and display utilizing the same" by Richard Hornblad et. al. of Wisconsin and assigned to DCI Marketing. The inventors describe a display device that uses flat continuous tape in the form of a Möbius strip.

- *U.S. Pat. 04599586* (1986) "Möbius capacitor" by Thomas Brown of New York. The inventor describes a capacitive enclosure in the shape of a Möbius strip, along with an electric element to measure voltage and phase differences of input signals or to act as a filter to attenuate current flow.

- *U.S. Pat. 04384717* (1983) "Möbius strip puzzle" by Daniel Morris of Washington. The inventor describes a puzzle that uses multiple Möbius strips linked in novel ways.

- *U.S. Pat. 04253836* (1981) "Möbius belt and method of making the same" by Joseph Miranti of Missouri and assigned to Dayco Corporation. Power transmission belts are made from spliceless Möbius strips.

- *U.S. Pat. 04189968* (1980) "Möbius strip bandsaw blade," by Joseph Miranti of Missouri and assigned to Dayco Corporation. The inventor describes a bandsaw with a blade in the shape of a Möbius strip.

- *U.S. Pat. 04058022* (1977) "Möbius drive belt fastener" by Harry Pickburn of New York. A Möbius drive belt and Möbius drive belt fastener are used for transmitting power between pulleys. The Möbius fastener allows the belt to be rotated in numerous ways.

- *U.S. Pat. 04042244* (1977) "Möbius toy" by Thomas Kakovitch of Maryland. This handheld toy challenges the manual dexterity and concentration of the user. The toy includes a Möbius ring formed from an elongated band having grooves in both "sides" to define a raceway for a rolling-ball playing piece. The band contains a hole, closed by a one-way door, to allow the ball to selectively visit either side of the puzzle.

- *U.S. Pat. 3,991,631* (1976) "Woven endless belt of a spliceless and Möbius strip construction" by J. Lehman Kapp of North Carolina. The inventor describes a Möbius belt used in manufacturing and material handling operations. The belt is characterized by having increased surface and edge wear potential.

- *U.S. Pat. 3,953,679* (1976) "Telephone answering device utilizing Möbius loop activating switch" by Neal Buglewicz of California and assigned to Phone-Mate. A telephone answering device uses a conventional tape recorder and an endless broadcast recording tape in the form of Möbius loop.

- *U.S. Pat. 3,758,981* (1973) "Möbius band type amusement device" by Richard Hlasnicek et. al. of Colorado. The inventors describe a toy that includes transparent tubing mounted on a Möbius strip. Steel balls travel in the tube. The inventors note, "Movement of the balls in the tubing may produce an audible sound that adds to the interest of the observer handling the device."

- *U.S. Pat. 3,648,407* (1972) "Dynamic Möbius band" by Jerome Pressman of Massachusetts. The inventor describes a Möbius band track along with a self-propelled vehicle that moves about the track "to demonstrate the one-sided topological characteristics of the surface."

- *U.S. Pat. 3,621,968* (1971) "Ribbon cartridge with Möbius loop in ribbon" by Nicholas Kondur, Jr. of Michigan and assigned to Burroughs Corporation. This inventor describes an inked Möbius ribbon that has double the effective length of a ribbon without a twist. The ribbon is moved by a drive roller on a printing machine.

Knot Patents: From Shoelaces to Surgery

Various knots have been invented and patented in a variety of fields. For example, figure 4.10 shows the patented "Partially tied surgical knot" (U.S. Pat. 5,893,592). According to the author, the knot is useful during "minimally invasive surgical procedures where access to the site is limited."

U.S. Pat. 5,997,051 describes a shoelace tying system for use with shoelaces in sneakers, shoes, or boots (figure 4.11). This 1999 patent, by

Paul and Majorie Kissner describes how to tie shoes that "resist inadvertently becoming untied . . . but can easily be untied when desired."

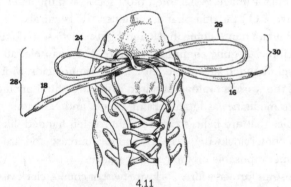

4.10
Partially tied surgical knot (U.S. Pat. 5,893,592).

4.11
U.S. Pat. 5,997,051 describes a shoelace tying system.

One wonders how much money is collected from people who tie these knots and thereby infringe on the patent.

Möbius and Knotty Chemistry

In the previous sections, I've emphasized the occurrence of the Möbius strip in visible objects; that is, objects we can touch, see, and feel, like Möbius conveyor belts and toys. In this section, we explore Möbius strips and trefoil knots on the molecular level, about which little was known until recently. Before giving concrete examples, let's examine *chirality* in chemistry. A chiral molecule is one that is not superimposable on its mirror image, in the same way that these two symbols, in a 2-D universe, cannot be made to superimpose no matter how we slide them around on the plane of this page:

More particularly, the two mirror forms of a chiral molecule—called enantiomorphs—cannot be mapped to each other by rotations and translations alone. Chirality also refers to the "handedness" of a molecule that is not symmetrical. As an analogy, your right hand and left hand are enantiomers: they are mirror images, and you cannot wear a left glove on your right hand. The arrangement of your thumb and fingers in three dimensions makes your right hand and your left hand different from each other. This handedness property is known as chirality, which is derived from the Greek word for hand.

Some objects are achiral because they have mirror images that *can* be superimposed upon each other. In other words, such an object is identical to its mirror image. A hammer, most socks, and the letter "I" are all achiral objects. On the other hand, the letter "R" is chiral.

Just as human hands come in left and right varieties, so do many molecules. Chirality is quite common in nature. For example, all creatures use only right-handed sugars and left-handed amino acids. More than 50 percent of the world's top one hundred drugs are chiral, including such well-known medicines as Lipitor, Paxil, Zoloft, and Nexium.

Molecules that are helices can be right- or left-handed, like a clockwise and counterclockwise seashell or spiral staircase and are therefore also nonsuperimposable on their mirror image.

Möbius strips formed with a clockwise twist or counterclockwise twist are enantiomers of each other. In the early 1980s, chemists were able to synthesize Möbius-band molecules with a carbon and oxygen backbone as schematically shown in figure 4.12. In 1982, David Walba, Rodney Richards, and R. Curtis Haltiwanger of the University of Colorado at Boulder discovered an efficient means for synthesizing the first molecular Möbius strip ever made by humans. The edge of the molecular band is traced out by chemical bonds, while the interior of the band is represented by a sequence of "rungs" formed by carbon double bonds. To create the structure, they started with a molecule shaped like a ladder with three rungs, each rung a carbon-carbon double bond (figure 4.13, top). During the chemical reaction, the ladder curved so that the ends could be joined. The two ends had the potential to unite in one of three ways, so that Mother Nature sometimes creates a circular loop–like a cylindrical strip–and at other times the loop will be a Möbius strip with either a clockwise or counterclockwise twist. In principle, more highly twisted molecules may result, but space-filling models indicate that such reactions are very unlikely. Figure 4.13 is a schematic representation of the parent molecule, diol ditosylate, that may produce the two molecules–the Möbius molecule (two variants on left) and the cylinder (right).

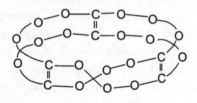

4.12

A Möbius band molecule. For diagrammatic simplicity, the O-O symbols are not actual bonds, because each has a -CH$_2$CH$_2$- group between them. The additional atoms are omitted in this figure to make it easier to visualize the Möbius configuration.

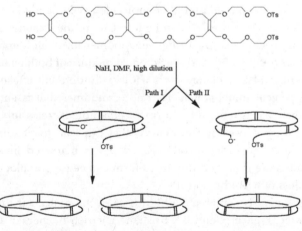

4.13

Creating a Möbius molecule. The ladderlike diol ditosylate (top) has ends that, when joined, produce roughly equal amounts of two molecules, the Möbius molecule (two variants on the left) and the cylinder (right). (Figure courtesy of David Walba.)

The properties of the ordinary paper Möbius strip are exhibited in Walba's microscopic biochemical one. Breaking the three rungs (carbon-carbon bonds) holding the two molecular edges together corresponds to cutting a Möbius strip along its middle, which produces one longer loop.

When divided this way, the Möbius molecule also becomes a single band with twice the circumference of the original.

The two different molecular Möbius strips, half twisted clockwise and counterclockwise, are not topologically equivalent because neither can be deformed into the other. If you looked at either strip in the mirror, the reflection of one looks like the other because the two strips are mirror images of each other.

Usually, chemists have a hard time controlling the amount of each mirror-image form, or enantiomers, in chemical reactions in drug preparations. For example, the drug thalidomide has a right-handed form that is useful in sedating pregnant women. However, the left-handed form causes birth defects. Tragically, many women took thalidomide in the 1960s for the sedative effect and gave birth to children with severe birth defects due to the presence of the left-handed version. However, even if women were to take only the right-handed form, birth defects would result because the enantiomers are converted to each other in the body. This means that if a woman is given either enantiomer, both isomers will be found in the blood.

In the case of the common pain reliever ibuprofen (found in Advil, Motrin, Nuprin, and Medipren), the molecule's right-handed form is one hundred times less powerful than its left. Due to the expense and difficulty involved in preparing a single-enantiomer form, all ibuprofen formulations currently marketed are an equal mixture of both enantiomers.

Other examples of enantiomers with vastly different effects include the drug penicillamine, which has one enantiomer that is antiarthritic and another that is toxic. One form of ethambutol treats tuberculosis, while the other causes optical neuritis that can lead to blindness. The Parkinson's disease drug levodopa (L-dopa) is marketed in an enantiomerically pure form because the D-form can cause granulocytopenia, that is, a loss of white blood cells.

Today, chemists can create all kinds of exotic molecular topologies. For example, French scientists Christiane Dietrich-Buchecker and Jean-Pierre Sauvage from the Université Louis Pasteur in Strasbourg have made molecular trefoil knots of various kinds.

Sauvage is a synthetic chemist who has been fascinated by the aesthetics of the molecules he creates and says that "the search for aesthetically attractive molecules has been a goal since the very origin of chemistry." In particular, the trefoil knot holds particular interest for him, as it has represented "continuity and eternity in early religious symbolism" and illuminates the art of many ancient civilizations.

In the last few years, Sauvage's group has paved the way for the preparation of knots constructed around transition metal ion templates. To create molecular knots, he uses two molecular "threads," which are interlaced on two transition metal centers, leading to a double helix. After cyclization and demetalation, a knotted system results. Several years of research were necessary for him and Dietrich-Buchecker to finally synthesize the first chemical trefoil knot using copper atoms as templates. Sauvage writes to me, "Christiane Dietrich-Buchecker and I were the first people to make a trefoil knot with molecules—at least, artificially, because nature has made molecular knots for millions of years, with DNA or proteins!" Figure 4.14 shows a schematic chemical diagram of a left-handed and right-handed knot created by Sauvage and colleagues.

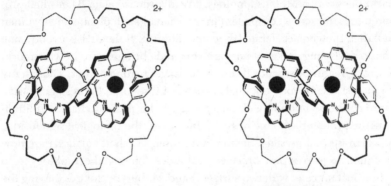

left-handed knot   right-handed knot

4.14

Chemical diagram of a left-handed and right-handed knot created by Jean-Pierre Sauvage and Christiane Dietrich-Buchecker.

In chapter 2, we discussed the work of British mathematical biologist William R. Taylor who developed an algorithm for detecting knots in protein backbones, and the various trefoil and other knotted proteins he discovered as a result of scanning protein structures stored in the Protein Data Bank. In 2002, researchers at the Argonne National Laboratory and the University of Toronto found a knot in a protein from the most ancient type of single-celled organism, an archaebacterium. Long ago, protein folding experts believed that forming a knot was beyond the ability of a protein. Today, protein chemists not only know that knots exist but they hypothesize that some of the discovered knots may stabilize the amino acid subunits of the protein. Biochemists hope that understanding the entire range of protein shapes will shed light on the process that proteins use to fold into three-dimensional structures from linear chains of amino acid subunits. If scientists can accurately predict the shape of a protein from the sequence of the gene that codes for it, they may be able to better understand diseases and develop new drugs that rely on a protein's 3-D shape, which controls the protein's function.

The archaebacterium knot discovered in the primitive organism in 2002 comes from *Methanobacterium thermoautotrophicum,* a primitive creature able to break down waste products and produce methane gas. Biochemists know which gene codes for the 268-amino acid protein, but they do not know the knotted protein's function.

A few remarkable proteins have recently been discovered to contain backbones with a Möbius topology. My favorite is kalata B1, a small protein that is the active component in the kalata-kalata plant, which women in Africa brew to accelerate labor in childbirth. Kalata B1 is twenty-nine amino acids long and contains three disulfide bridges, two of which form a ring. The third descends through the ring, creating a knot. Though my topologist friends may not qualify this as a true knot, the resultant structure cannot be undone without damaging the protein chain. Additionally, the protein backbone has a Möbius twist, the biological significance of which is still a mystery. Small plant proteins such as kalata B1 are now classified as two types: bracelet cyclotides (having the topology of a cylindrical bracelet with two surfaces) and Möbius cyclotides (having the topology of a Möbius strip with one surface). Kalata B1 is a Möbius cyclotide. Because kalata B1 has insecticidal and antimicrobial properties, scientists are now considering its use to protect crops through bioengineering. Professor David Craik from the University of Queensland suggests that kalata B1 could be inserted into a cotton plant's genes to protect the plant from the ravages of certain caterpillars, thereby removing the need to use chemical sprays that are of environmental concern. Moreover, because the protein is very stable and resistant to attack from digestive enzymes in the human body, perhaps kalata-like proteins will someday be used as a framework for new drugs that can be taken orally because they are not broken down so quickly in the stomach.

Moving to much larger Möbius objects in the field of chemistry, in 2002, researchers at Hokkaido University in Japan described a Möbius loop formed by crystals of a compound of niobium and selenium, $NbSe_3$. They were surprised that a crystalline ribbon should adopt the Möbius topology in view of the crystal's inherent rigidity, which would be expected to prevent it from either bending or twisting. $NbSe_3$ is an inorganic conductor, and the Möbius strip is a single $NbSe_3$ crystal. To produce the shapes, they soaked a mixture of Se and Nb powder at 740°C in an evacuated quartz tube for about a day. Scanning electron microscopic images reveal Möbius crystals typically around 50 micrometers in diameter and less than 1 micrometer in width. The Japanese researchers believe that their Möbius crystals offer "a new route to exploring topological effects in quantum mechanics as well as to the construction of new devices."

Other scientists also study the properties of real and hypothetical molecules that have interesting twists and turns. Figure 4.15 is a hypothetical Möbius molecule studied by European chemists Sonsoles Martín-Santamaría and Henry S. Rzepa. These chemists study the properties of Möbius strips formed by imparting one, two, or three twists to various lengths of molecules known as cyclacenes. This molecule is special in that it is an *aromatic* molecule—namely a molecule in which electrons are free to cycle around circular arrangements of atoms, which are connected by bonds that are a hybrid of a single bond and a double bond.

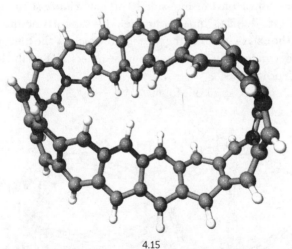

4.15
A molecular Möbius band courtesy of Henry S. Rzepa.

Dr. Rzepa's work was theoretical, but in December, 2003, German scientists reported the actual synthesis of the world's first Möbius aromatic hydrocarbon molecule. The German researchers discovered an ingenious method for joining two eight-carbon aromatic molecules to form a sixteen-carbon non-orientable molecule. To create the molecule, researchers irradiated the reaction mixture for four hours with a mercury lamp. The Möbius molecules that resulted formed red crystals, whereas non-Möbius molecules form clear crystals.

Holiday Möbius Train

I finished writing this chapter on Christmas Day and would like to conclude with a Möbius patent from China for a Möbius strip train track suitable for winding around Christmas trees. In particular, U.S. Pat. 5,678,489, "Electrically-operated moving body traveling on a rail capable of explaining free quadrants described in the Möbius theorem" was issued in 1997 to Xian Wang of Changsha, China and assigned to Studio Eluceo Ltd. and Jya Cheng Enterprise Co. Ltd. The inventor describes a delightful electric train traveling on a Möbius train track with various supports. Two parallel metal tracks are used so that the train can make contact on "upper" and "lower" surfaces of the tracks. A controller is used to adjust the current to the tracks and power the train, which has rollers made of permanent magnetic material that are attracted to the tracks.

The patent gives further details on how the train is held on the track when upside-down on the one-sided surface. Figure 4.16 shows an example of the Möbius Chinese train held by supports on the ground. Figure 4.17 shows a figure from the patent showing how the train track can be used to decorate a Christmas tree as it meanders around the branches and through ornaments.

Merry Christmas, spirit of Möbius.

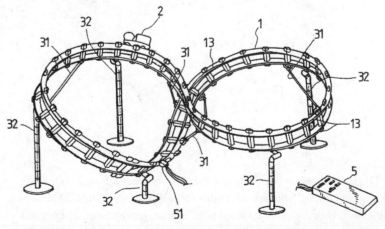

4.16

U.S. Pat. 5,678,489 issued in 1997 for an electric Möbius track and train set.

4.17
Electric Möbius train track meandering about a Christmas tree.

◉ Graph Theory Puzzle Involving Animals

In the next chapter, we'll discuss graph theory, which often deals with different ways of connecting objects. To whet your appetite, here's a problem that many of my colleagues find impossible to solve, at least at first glance.

Noah has unloaded his ark, and the animals have scattered in their haste for freedom. His job is now to unite male and female of the same species. In figure 4.18, the rabbit, horse, and elk at the bottom are all touching the south wall of a very large fenced enclosure. Three other animals are positioned so that the horse is touching the north wall. Is it possible to connect animals of the same kind using lines you draw along the ground? The lines may not cross or touch the enclosure walls. (In other words, you must try to draw a line from the rabbit to the rabbit, the horse to the horse, and the elk to the elk.) The paths you draw may be curvy, but they cannot touch or cross each other, nor can they go "through" the animals or touch the lake. If you can solve this problem within five minutes, Noah will give you a golden Möbius strip worth $1 million in today's currency. (Turn to the solutions section for an answer.)

4.18
A simple graph? Draw lines to connect animal pairs without crossing lines and without drawing a line through the animals or the boundaries.

Möbius Strip in Fashion and Hairstyle

🐾 *United Nude, an innovative shoe-design company, looks to graphic designers to design materials for its new shoe collections. United Nude's unique design of the Möbius shoe, shaped from one continuous piece of material, a Möbius strip, has been inspired by architect Mies van der Rohe's iconic Barcelona chair.*
—Barbara Wentzel, "Pushing the Boundaries," World Textile Publications

🐾 *Carneval Möbius Stole—$48.00. This is so much fun to wear! We used "Carneval," a wonderful soft cotton & rayon luscious, lofty yarn from Muench Yarns, and knit every row into a Möbius stole that can dress up any summer outfit. You can move in this and the stole stays in place. We highly recommend this yarn for its quality, drape and natural fibers. Works well in large size. A Knittingbag design.*
—Knittingbag.com Catalogue, Knitting Bag, LLC

🐾 *Is Donald Trump's hair brushed forwards or backwards? (Seriously, it's like an infinity pool now, like a furry Möbius strip with no beginning or end, just flow.)*
—Whitney Pastorek, "Donald toys with 18 new victims,"
Entertainment Weekly, September 2004

STRANGE ADVENTURES IN TOPOLOGY AND BEYOND

Möbius's working habits demonstrate most clearly how marvelous discoveries can be made by patiently building on the simplest cases and always working fully through the special (and seemingly least interesting) examples. We can all wait for genius to strike, but patient work also brings its rewards.

–Jeremy Gray, "Möbius's Geometrical Mechanics,"
in Möbius and His Band

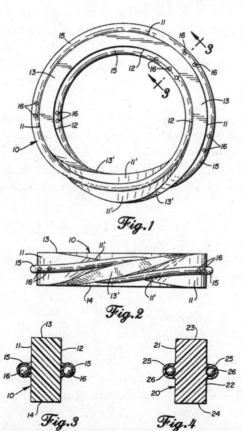

Fig.1

Fig.2

Fig.3   Fig.4

Möbius's mind ranged far and wide—from one-sided surfaces to strange integer functions such as the one we today call the "Möbius function" in his honor. In this chapter we'll discuss several of his mathematical discoveries. We'll also contemplate topology and higher dimensions, which provide background for the next chapter on the shape of the universe and the possible existence of other realities.

The Endurance of Mathematics

Benoît Mandelbrot, the father of fractal geometry, once said that a mathematical topic dating from 150 years ago is old "but not dead and dried to dust." In contrast, he observes that this is very different in a field like physics, "where something that is 100 years old but not in textbooks is, for all practical purposes, dead." This endurance of mathematics is never more apparent than with Möbius's work. His one-sided surfaces and bizarre functions are still contemplated with zeal today, with new discoveries and intuitions emerging frequently with the aid of computers, and with implications beyond mathematics, such as in the behavior of subatomic particles, the shape of space, and the genesis of our universe.

Parameterization

Mathematicians sometimes use parametric equations to represent sophisticated geometrical shapes. Parameterizations are sets of equations that express a set of quantities as functions of a number of independent variables. Perhaps the most famous examples are the equations for the circle. In the usual Cartesian coordinates, we have the standard equation of a circle:

$$x^2 + y^2 = r^2$$

where r is the radius of the circle. We can also define a circle in terms of parametric equations:

$$x = r \cos(t)$$
$$y = r \sin(t)$$

where $0 < t \leq 360$ degrees or $0 < t \leq 2\pi$ radians. To create a graph, computer programmers increment the value for t and connect the resultant (x, y) points. The smaller the increments in t, the smoother the resultant representation of the circle.

Mathematicians and computer artists often resort to parametric representations because certain geometric forms are very difficult to describe as a single equation, as we could for a circle. For example, one parameterization of a circular helix is: $x = a \sin(t)$, $y = a \cos(t)$, and $z = at/(2\pi c)$ where a and c are constants. Try $a = 0.5$, $c = 5.0$, and $0 < t < 10\pi$. A plot of this circular helix curve resembles a wire spring. To draw a conical helix, try $x = a \times z \times \sin(t)$, $y = a \times z \cos(t)$, and $z = t/(2\pi c)$ where a and c are constants (figure 5.1). Conical helices are used today in certain kinds of antennas.

5.1 Conical helix. (Rendering by Jos Leys.)

One of my favorite parameterizations is represented by spherical Lissajous curves generated from

$$x = r \sin(\theta t) \cos(\phi t), \; y = r \sin(\theta t) \sin(\phi t), \; z = r \cos(\theta t)$$

which I used to create the artwork in figure 5.2. You can try simple ratios of θ/ϕ such as $1/2$ or $1/3$ to produce visually interesting results.

5.2
Spherical Lissajous curve.

In fact, even if we restrict ourselves to the plane, incredible beauty can be found in a variety of algebraic and transcendental curves. Many of these curves express beauty in their symmetry, leaves and lobes, and asymptotic behavior. Butterfly curves, developed by Temple Fay at the University of Southern Mississippi, are one such class of beautiful, intricate shapes (figure 5.3). The equation for the butterfly curve can be expressed in polar coordinates by

$$\rho = e^{\cos\theta} - 2\cos4\theta + \sin^5(\theta/12)$$

This formula describes the trajectory of a point as it traces out the butterfly's body. ρ is the radial distance of the point to the origin.

5.3
Butterfly curves defined by a simple formula.

With this introduction to parametric equations, we can now ponder equations for the Möbius strip. One typical parameterization is:

$$x(u,v) = \left(1 + \frac{v}{2}\cos\frac{u}{2}\right)\cos(u)$$
$$y(u,v) = \left(1 + \frac{v}{2}\cos\frac{u}{2}\right)\sin(u)$$
$$z(u,v) = \frac{v}{2}\sin\frac{u}{2}$$
$$(0 < u \le 2\pi; -1 < v < 1)$$

This creates a Möbius strip of width 1 and radius 1 that is centered at (0, 0, 0). The parameter u runs around the strip like a racer along a racetrack. The parameter v moves from one edge to the other.

It's also common to represent a Möbius strip in cylindrical polar coordinates (r, θ, z) by $\log(r)\sin(\theta/2) = z\cos(\theta/2)$.

Paradromic Rings

Researchers from the late 1800s to the present day have catalogued the effects of giving extra twists to pieces of paper before joining the ends to form Möbiuslike strips. The surprising results are called *paradromic rings*. Some of the possible paradromic ring structures are listed in the following table.

Half Twists	Cuts	Apparent Pieces	Actual Result
1	1	2	1 band, length 2
1	1	3	1 band, length 2 1 Möbius strip, length 1
1	2	4	2 bands, length 2
1	2	5	2 bands, length 2 1 Möbius strip, length 1
1	3	6	3 bands, length 2
1	3	7	3 bands, length 2 1 Möbius strip, length 1
2	1	2	2 bands, length 1
2	2	3	3 bands, length 1
2	3	4	4 bands, length 1

For example, the first row of the table corresponds to a cut down the center of a Möbius strip, as in figure 1.1. When your cut is complete, you have two "apparent strips" or "divisions." However, when you then stretch out the result, you find you have only one piece, as shown in the final column of the table. In the second row, we cut the Möbius strip a third of the way from an edge, a remarkable experiment we discussed in chapter 2. In this case, we produce three apparent pieces, which, when stretched out, reveal themselves to actually be one Möbius strip and one loop (two pieces). In the third and fourth rows, we are allowed two cuts; depending on where the cuts are made, there will be either four or five apparent pieces when we are finished cutting, but when we stretch them out, we find two or three loops, respectively.

Notice that nowhere does the table tell us *where* to make the cuts on the strip. It simply shows various possibilities. If you make a Möbius strip, you'll find that it really doesn't matter where you make the cut indicated in the table's second column, as long as your cut is not in the center. The first two rows describe nearly the same situation—the one cut in the first row is a cut along the center; the second row describes an off-center cut. A similar logic applies to other rows in the table.

Adventures in Topology

According to Norman Biggs, author of "The Development of Topology" in *Möbius and His Band*, Möbius probably did not think of himself as a topologist because there was no general area of mathematics called topology during his lifetime. Nevertheless, his ideas, papers, and diagrams have had a profound influence on the development of topology.

The river of topology has its source in Leonhard Euler (1707–1783), the Swiss mathematician and physicist who is often considered, together with Gauss, to be the greatest mathematician who ever lived. One of Euler's interests centered on ordinary shapes with corners, faces, and edges. Years later, Möbius became fascinated by Euler's work that established relationships between the number of edges, vertices, and faces of a simply connected polyhedron. To understand this area of geometry, let us imagine that Möbius lived in a house in Schulpforta that resembles the shape in figure 5.4.

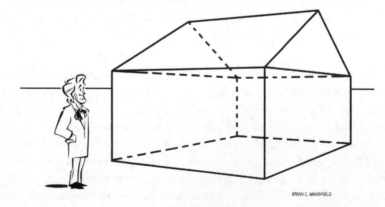

BRIAN C. MANSFIELD

5.4
The house that Möbius built.

For simplicity, we will assume that his house has no windows and doors—not a very practical house but a perfect model for us to use when studying topology. Möbius's simple house has 10 vertices (V), 17 edges (E) and 9 faces (F). If you don't see this, remember that a vertex is a corner, an edge is where two walls meet, and a face is either a wall, part of a roof, or the floor. Euler observed that for a house like Möbius's, we have the relation:

$$V - E + F = 2.$$

For example, for this house, we have $10 - 17 + 9 = 2$. Antoine-Jean Lhuilier (1759–1840) wondered if Euler's formula still would work with more complex shapes, such as a simple house with a courtyard (figure 5.5).

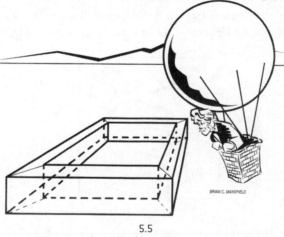

BRIAN C. MANSFIELD

5.5
The house that Möbius built, with central courtyard.

This house has 16 vertices, 32 edges, and 16 faces. Inserting these values into Euler's formula we get

$$V - E + F = 0$$

Uh-oh! This shows us that Euler's formula does not work once we add windows, courtyards, and doors to Möbius's simple house. But by making a quick fix, Lhuilier discovered a more general formula:

$$V - E + F = 2 - 2G$$

where G is the number of holes in an object. For example, a house with two separate interior courtyards, like a brick with two square holes drilled all the way through, has $G = 2$. It turns out that $V - E + F = 2 - 2G$ is true for a wide variety of shapes. (Some complexities arise when we want to define the "number of holes" because holes may join and coalesce in curious ways, like tunnels in an ant colony, or like the hole within a hole through a hole in the introduction.)

Notice that the same formulas apply to flat maps such as the one in figure 5.6, for which we have $V - E + R = 2$. Here, R is the number of regions, including the outer boundary of the planar map. In figure 5.6, $V = 15$, $E = 23$, and $R = 10$, and the formula holds. The formula fails the moment we add a disconnected oval region, corresponding to a hole or lake, which represents a vertex connected to itself. However, if we connect the hole to one of the edges with a line segment, as indicated by the dashed line in figure 5.7, the formula becomes valid again. By adding the line, we have also added another vertex (at the far point of the line) and created two additional edges.

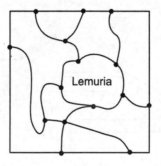

5.6
Lemuria map can be used to demonstrate Euler's formula, $V - E + R = 2$.

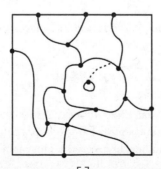

5.7
Map containing an island nation.

While on the subject of maps, note that the *chromatic number* of a surface describes the maximum number of regions that can be drawn on the surface so that each region is given a different color, yet each color will border every other color. For example, the chromatic number for a square sheet is four (figure 5.8).

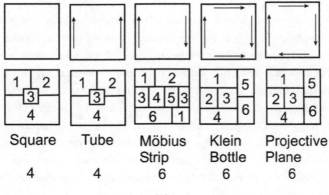

5.8
Map coloring on a variety of surfaces. The square diagrams at the top show how edges join in each model. The second row of squares shows one way the surface can be mapped to accommodate the maximum number of colors. Chromatic numbers are at the bottom.

To understand figure 5.8, note that the square diagrams at the top show how edges are joined in each model. Sides with arrows are "sticky" and connect to each other with the directions of their arrows coinciding. The second row of squares shows one way that the surface can be mapped to accommodate the maximum number of colors, which are represented by different numbers. If you were to actually color the regions different colors, you would color regions of both sides of the paper (as though the paper were translucent) because you must think of the sheet as having zero thickness. Chromatic numbers are along the bottom of figure 5.8.

Today, we know that a square, cylinder, and sphere have a chromatic number of 4. A Möbius strip, Klein bottle, and projective plane have a chromatic number of 6. (We'll discuss Klein bottles and projective planes later in this chapter and in the next chapter.) A torus, which you can think of as the surface of a doughnut, has a chromatic number of 7. This means that on a Möbius strip, six colors are needed to ensure that no bordering areas on any map will be colored the same. (Although we can find an example of a map requiring six colors for the Möbius strip, this doesn't mean that it is necessary that *every* map require six colors on a Möbius strip.)

Using the metaphor of a geopolitical map, the chromatic number of a surface is the minimum number of colors needed to properly color any map on the given surface so that countries with common borders get distinct colors. Thus, if you draw a complicated map, like a map of the United States, on a Möbius strip, then it is *possible* to color it with *at most* six different colors so that no two adjacent regions have the same color.

Mapmakers have known for centuries that just four colors are sufficient for coloring any map drawn on a plane—so that no two distinct regions that share a common edge are the same color, although two regions can share a common vertex and have the same color. Some planar maps require fewer colors, but all maps can certainly be done with four. Four colors are sufficient for maps drawn on spheres and cylinders, but seven colors are sufficient to paint any map on a torus.

Although no one had ever found a map on a plane that needs more than four colors, for about a century mathematicians tried in vain to prove this apparently simple theory. Finally, in 1976, mathematicians succeeded in proving the four-color theorem with the help of a computer, making it the first problem in pure mathematics to use a computer to produce an essential component for the proof.

Today, computers are playing increasing roles in mathematics, helping mathematicians verify proofs so complex that they sometimes defy human comprehension. The four-color theorem is one example. Another is the classification of finite simple groups, described in a multiauthor project of ten thousand pages. As Dana Mackenzie points out in *Science* ["What in the Name of Euclid is Going on Here?"], the traditional people-centered methods for ensuring that a proof is correct breaks down when a paper reaches thousands of pages. With respect to the four-color theorem, graph theorist John Robin Wilson notes, "Mathematicians over 40 years old couldn't be convinced that a proof by computer was correct, and those under 40 couldn't be convinced that a proof with 700 pages of hand calculations was correct." A "streamlined" proof of the four-color theorem was published in 1995. Even with this more compact approach, a computer was required to check more than a billion different maps, something that would take a human mathematician many lifetimes.

I point out in my book *A Passion for Mathematics* that we live in an age where even simple computer tools like spreadsheets give modern mathematicians power that Heisenberg, Einstein, and Newton would have lusted after. As just one example, in the late 1990s, computer programs designed by David Bailey and Helaman Ferguson helped produce new formulas that related pi to log 5 and two other constants. Erica Klarreich reports in the April 24, 2004, *Science News*, once the computer had produced the formula, proving that it was correct was extremely easy. Often, simply *knowing* the answer is the largest hurdle to overcome when formulating a proof.

The four-color theorem has even fascinated science fiction authors. Martin Gardner, in "The Island of Five Colors," tackles the four-color theorem, which was unproved at the time he wrote the story in 1952. Gardner's tale is loaded with geometrical musings and even mentions the correct chromatic numbers for the torus, Möbius strip, and Klein bottle. The protagonist also alludes to many exotic surfaces like the cross-cap (discussed later in this chapter), a Tuckerman strip (a Möbius strip with an edge in the form a triangle), and the two-layer sandwich Möbius strip that we discussed in chapter 2. In the story, we learn about prior attempts to prove the four-color theorem and explore a fictional counterexample in the form of a mysterious African island divided into five simply connected districts—each of which borders the other and the ocean. The protagonist is so confused by the remarkable island's arrangement of districts, which seem to contradict the four-color theorem, that he paints the island's districts with five colors, red, blue, green, yellow, and purple, and then has a friend take a photo from the air to help understand how the paradoxical district configuration could exist. In particular, the protagonist buys twenty thousand gallons of water-based paint and sprays the five districts with spots of color twenty feet in diameter at intervals of fifty yards. On the ground, it's difficult to tell what the district shapes are, but he hopes an aerial view will make this clear. Alas, the aerial photos do not develop properly, and he never understands the mysterious relationships between the shapes of the districts. The story ends with a math professor on the island who is suddenly yanked into a Klein bottle by a gigantic insect! Our protagonist looks into the bottle's opening but sees only swirling fog and feels an icy upward rush of air. He yells the professor's name, but only hears faint echoes and faraway voices in a strange language. What a story!

As an April Fools' joke, I once told friends that I had made the stunning discovery of an unusual configuration of hypothetical countries that also required five colors to color the regions, so that two regions with a common edge did not have the same color (figure 5.9). Some friends slaved at coloring my map for a long time and triumphantly showed me their results, before realizing I was just joking.

5.9
Da Vinci code map. How many colors do you need to color this unusual configuration of hypothetical countries so that regions with a common edge do not have the same color?

Robin Wilson, author of *Four Colors Suffice*, makes the surprising observation that the four-color problem has been of little importance to mapmakers and cartographers. As evidence, mathematical historian Kenneth May observed that a sample of atlases in the large collection of the Library of Congress indicates "no tendency to minimize the number of colors used," and "maps utilizing only four colors are rare." Moreover, books on cartography and the history of mapmaking do not mention the four-color property.

Möbius's Triangulated Band
This introduction to some fundamental concepts of basic topology leads us to Möbius's band and Möbius's thoughts when he made his remarkable finding. Möbius, like Antoine-Jean Lhuilier, wondered what geometry and topology would be like for objects that are more unusual than the simple Möbius house (figure 5.4) that started our discussion. To facilitate his

studies of one-sided surfaces, Möbius constructed surfaces from flat triangular pieces. For example, a faceted Möbius band can be made from a number of triangles as shown in figure 5.10.

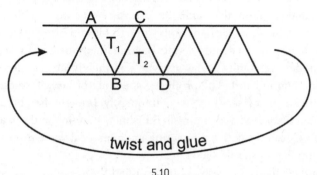

5.10
The Möbius band represented as a surface of triangular facets.

Möbius explained one-sidedness in terms of the way the triangular facets of an object fit together. We can understand Möbius's thinking by defining the difference between clockwise and counterclockwise rotations of nearby triangles. For example, in triangle 1 in figure 5.10, we can define the order A-B-C to be counterclockwise. In triangle 2, let's start from the strip's top edge as we did for triangle 1. For triangle 2, C-B-D is counterclockwise. Notice that shared edges B-C in triangle 1 and triangle 2 have a different order with respect to the rotations of the triangles. Triangles are called "compatible" when their shared edge is oriented in these opposite senses. However, at the location where we join the edges of the strip to form a Möbius band, the adjacent triangles will not be compatible. Thus, just like Euler, Möbius focused on edges and vertices, and he extended previous work to study Euler-like characteristics in one-sided objects.

Dr. Johann Listing and Homeomorphisms
In 1858, German mathematician Johann Benedict Listing (1808–1882) codiscovered the object we now call the Möbius strip. Listing is also often regarded as the founder of topology because in 1847 he wrote a book titled *Vortstudien zur Topologie,* and he had coined the word topology at least a decade earlier. Many of his topological ideas were probably stimulated by Gauss, with whom Listing first studied in 1829. Listing's 1861 book *Der Census räumlicher Complexe oder Verallgemeinerung des Euler'schen*

Satzes von den Polyëdern (*The Census of Spatial Complexes or Generalizations of Euler's Theorem on Polyhedra*) contains a description of the Möbius band. Some have speculated that Gauss may have given both Möbius and Listing the idea of a Möbius band, but this is mere speculation.

Incidentally, about the same time Listing was contemplating the Möbius band, he was near bankruptcy due to his and his wife's mismanaging of family finances and living beyond their means. Listing borrowed money frequently while his wife continually abused credit and often ended up in court. The profligate spending of his wife, along with her mistreatment of servants, who frequently brought her before the magistrates, is said to have diminished Listing's standing in the academic community. As a result, his pioneering mathematical work received less recognition than it should have.

Among his many interests, Listing studied the value of $V - E + F$ for polyhedra, which today is known as the polyhedron's Euler number. As we have discussed, the number is constant for solids that have the same number of holes and pieces; in other words, for solids that are related by transformations called *homeomorphisms* that involve no cutting and tunneling.

Two geometrical objects are called homeomorphic if the first can be deformed into the second by stretching and bending. (Technically, cutting is sometimes permitted, but only if the two parts are later glued back together along exactly the same cut, and neighboring points before the cut are neighbors after the cut.) For example, a square and a circle are homeomorphic. A hollow sphere containing a smaller solid ball is homeomorphic to a hollow sphere with a solid ball outside of it. You can cut the outer sphere, move the inner ball through the cut to the outside, and then glue the sphere together along exactly the same cut. Or you can translate one sphere in the fourth dimension and return it to the third dimension. If two objects are homeomorphic, one can find a continuous function that maps points from the first object to corresponding points of the second object and vice versa. Such a function is called a homeomorphism, and it must map points in the first object that are "close together" to points in the second object that are "close together." Topology is the study of those properties of objects that do not change when homeomorphisms are applied.

The classic example of a homemorphism is the transformation of a doughnut into a coffee cup. We can deform a very malleable, rubberlike doughnut into the shape of a coffee cup without any cutting or pasting. The hole in the doughnut simply becomes the handle, with its hole, for the coffee cup. On the other hand, the surface of a doughnut (called a torus) is not the same as the surface of a solid ball (called a sphere). There is no way to morph one into the other without cutting and pasting in ways that are not permitted with a homeomorphism.

Operations requiring tunneling are not homeomorphisms, and cutting a faceted object may change the value of $V - E - F$. Interestingly, homeomorphisms ignore the space in which surfaces are embedded, so deformations are permitted to be completed in a higher dimensional space. This means that mirror images are homeomorphic because they can be transformed into each other by rotation in a higher dimension. (If this is not clear to you, it will become obvious in the next chapter when we slide 2-D congruent blobs on a plane; these blobs are not superimposable unless we lift one out of the plane and flip it over.) Thus, two Möbius bands that are mirror images of each other, because they are twisted in opposite directions, are topologically identical.

All Möbius strips with an odd number of half twists are homeomorphic to each other. And all strips with an even number of half twists are homeomorphic to each other. But a strip with an even number of half twists is *not* homeomorphic to one with an odd number of half twists. Also, a Möbius strip essentially comes in two forms: the right-handed and left-handed, which can be turned into each other only if we could rotate the strip in the fourth dimension.

The property of being one-sided or two-sided is a topological invariant; however much we bend and stretch the band, it continues to have only one side. This means that a one-sided surface like a Möbius strip cannot be turned into a two-sided one by topological transformations. As I mentioned, with topological transformations we may, *in principle*, cut the strip, twist it, and even tie it into knots, but we must paste the strip back together in such a way that adjacent points before the cut are still adjacent. (This kind of pasting is why a trefoil knot and a circle are homeomorphic.) However, this constraint prevents us from adding or subtracting an odd quantity of half twists to strips. Today, topology and Möbius bands have important consequences for physics, cosmology, and mechanics.

Of course, a Möbius band can't really be transformed into its mirror image or into a band with three half twists in our universe. However, if it floated in 4-D space, we could deform it in a higher dimension and return it to 3-space as a loop with any odd number of half twists or either handedness. Even a band with no twists (like a cylinder) could, in theory, be lifted into 4-space, twisted by a higher-dimensional alien, and dropped back into our space with any even number of half twists of either handedness. However, the alien cannot change a standard cylindrical loop into a Möbius strip because the cylinder has two edges and the Möbius band has only one. To change two edges into one, the alien would have to break both of them and join them to each other's breaks.

Möbius gave a lot of thought to these weird higher-dimensional rotations. He wrote in *Werke* that a solid could be rotated into its mirror image if we were "able to let one system make a half revolution in a space of four dimensions. But since such a space cannot be conceived, this coincidence is impossible in this case."

Ghosts, Möbius Strips, and the Fourth Dimension

When I talk about the fourth dimension, I am referring to a spatial dimension that corresponds to a direction different from all the directions in our world. Students usually ask, "Isn't time the fourth dimension?" Time is *one* example of a fourth dimension, but there are others. Parallel universes may even exist besides our own in some ghostly manner, and these might be called other dimensions. But in this section, I'm interested in a fourth *spatial* dimension—one that exists in a direction different from up and down, back and front, right and left.

Look at the ceiling of your room. From the corner of the room radiate three lines, each of which is the meeting place of a pair of walls. Each line is perpendicular to the other two lines. Can you imagine a fourth line that is perpendicular to the three lines? If you are like most people, the answer is a resounding no. But this is what mathematics and physics require in setting up a mental construct involving four-dimensional space.

What does it mean for objects to exist in a fourth dimension? The scientific concept of a fourth dimension is essentially a modern idea, dating back to the 1800s. However, the philosopher Immanuel Kant (1724–1804) considered some of the spiritual aspects of a fourth dimension:

A science of all these possible kinds of space would undoubtedly be the highest enterprise which a finite understanding could undertake in the field of geometry. . . . If it is possible that there could be regions with other dimensions, it is very likely that a God had somewhere brought them into being. Such higher spaces would not belong to our world, but form separate worlds.

We've discussed the idea of aliens manipulating Möbius strips in higher dimensions, and this reminds me of the nineteenth-century astronomer

Johann Carl Friedrich Zöllner (1834–1882), who promoted the idea of ghosts from the fourth dimension. He was an astronomy professor at the University of Leipzig–Möbius's alma mater–and worked with the American medium Henry Slade. Zöllner performed various tests on Slade to determine if Slade was a charlatan or could really access higher dimensions. For example, Zöllner gave Slade a string held together as a loop. The two loose ends were held together using some sealing wax. Amazingly, Slade seemed to be able to tie knots in the string, something that should not be possible in a loop of string. Of course, Slade probably cheated and undid the wax, but if he could tie knots in the sealed string, it would suggest the existence of a fourth dimension. Let me explain why.

Imagine that we are together, and I hand you a string with a piece of wax sealing the two ends, as in figure 5.11. The circle with the "Z" represents the sealing wax that Zöllner applied. A four-dimensional being could move a piece of the string in a fourth dimension out of our space. This would be like cutting the string in the sense that the string could be moved through itself to form a knot. Once the string is oriented correctly, you could move it back "down" into our space, and a knot would be tied without moving the ends of the string. Similarly, a 4-D being could deform a Möbius band in a higher dimension and return it to 3-space as a loop with any odd number of half twists or either handedness. However, the being could not change a band with an odd number of half twists to a band with an even number without tearing the band because this conversion would change the strip from having one edge to having two edges. The edge would have to be broken in two places, and the ends of each of the two resulting pieces would have to be joined to make two edges.

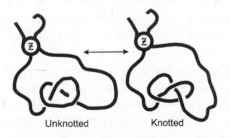

Unknotted   Knotted

5.11

A four-dimensional being would be the ultimate magician and could knot or unknot a string by temporarily lifting it into the fourth dimension. On the left is a string before it has been knotted. (Zöllner tried to transform the left string into the right without breaking the wax circle at the top.)

Let's think about knots in different dimensions. A string can't be knotted in two-dimensional space no matter how hard you try (figure 5.12). For creatures confined to living in a plane, there's no way a line can cross over itself. In fact, a string or a line can only be knotted in 3-D space. And any knot you tied in 3-D space will not stay tied in 4-D space because the additional degree of freedom will cause a knot to slip through itself.

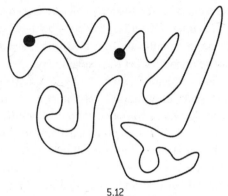

5.12
A string cannot be knotted in a two-dimensional space.

By analogy, in 4-D space a creature can knot a plane (surface), but this plane won't stay knotted in 5-D space. And the knotted plane cannot be formed in 3-D space. Of course, the idea of knotting a plane is probably confusing to most readers. Imagine a knotted line, and then move it "up" into the fourth dimension. The trail it traces will be a knotted plane. It never intersects itself. Of course, if we simply leave a trail in 3-space as we move a knot, it will intersect itself, but since this "up" is perpendicular to all directions in our space, the 4-D knotted plane will not intersect itself.

Zöllner devised three tests to determine if Slade could use the fourth dimension to perform miracles. First, he gave Slade two oak rings that were to be interlocked without breaking them. Second, he gave Slade a snail shell to see if a clockwise spiral could be turned into a counter-clockwise spiral, and vice versa. Third, he gave Slade a rubber band and asked him to place a knot in one strand of the band. (To be precise, it was a band made from dried gut, but you get the idea.)

Alas, Slade could not pass this difficult series of tests. Nevertheless, the idea of a fourth dimension continued to amuse and fascinate laypeople and scientists. "A higher world is not only possible, but probable," writes Alfred Taylor Schofield in his 1888 book *Another World.*

"Such a world may be considered as a world of four dimensions. Nothing prevents the spiritual world and its beings, and heaven and hell, being by our very side."

Zöllner was almost completely discredited because of his association with spiritualism. However, he was correct to suggest that anyone with access to higher dimensions would be able to perform feats impossible for creatures constrained to a 3-D world. He suggested several experiments that would demonstrate his hypothesis; for example, linking solid rings without first cutting them apart, or removing objects from secured boxes. If Slade could interconnect two separate unbroken wooden rings, Zöllner believed it would "represent a miracle, that is, a phenomenon which our conceptions heretofore of physical and organic processes would be absolutely incompetent to explain." Perhaps the hardest test to pass involved reversing the molecular structure of dextrotartaric acid so that it would rotate a plane of polarized light left instead of right. Although Slade never quite performed the stated tasks, he always managed to come up with sufficiently similar evidence to convince Zöllner, and these experiences became the primary basis of Zöllner's *Transcendental Physics*. This work, and the claims of other spiritualists, actually had some scientific value because they touched off a lively debate within the British scientific community.

Turning Spheres and Doughnuts Inside Out

The Möbius strip is one of a wide variety of exotic geometrical forms that play important roles in topology. Some topological transformations are easy to visualize. Most of us can imagine stretching a coffee cup into a doughnut, but topology also deals with many nonintuitive transformations. For example, for many years topologists knew that it was theoretically possible to turn a sphere inside out, yet they didn't have the slightest idea how to do it. When researchers began to use computer graphics, mathematician and graphics expert Nelson Max of the Lawrence Livermore National Laboratory produced an exciting animated film finally illustrating the transformation of the sphere. Max's 1977 movie was based on the 1967 sphere eversion work of Bernard Morin, a blind topologist at the Louis Pasteur University in Strasbourg, France. To create the film, Max started with a set of coordinates obtained from wire mesh models depicting eleven stages in the transformation.

Today, you can order the film "Turning a Sphere Inside Out" on the web from A. K. Peters. The animation begins with a discussion of the sphere eversion problem and focuses on how this can be done by passing the surface through itself without making any holes or creases. Mathematicians had believed that the problem was unsolvable until around 1958 when Stephen Smale, now at the University of California, Berkeley, proved otherwise. However, no one could visualize the motion without the graphics. The steps required to turn a *torus* inside out are also quite difficult to visualize.

When we discuss the eversion of a sphere, we're not talking about turning a beach ball inside out by pulling the deflated ball through its opening and then inflating it again. Instead, we are referring to a sphere with no orifice. Mathematicians try to visualize a sphere made out of a thin membrane that can stretch and even pass through itself without ripping or developing a sharp kink or crease. The task of avoiding such sharp creases makes the mathematical sphere eversion so difficult.

It turns out there are several ways to accomplish the task, and in the late 1990s mathematicians went a step further and discovered a geometrically optimal path—one that minimizes the energy needed to contort the sphere through its transformation. This optimal sphere eversion, or *optiverse*, is now the star of a colorful computer graphics movie titled *The Optiverse*, produced by mathematician John M. Sullivan of the University of Illinois at Urbana-Champaign and his Illinois colleagues George K. Francis and Stuart Levy. However elegant this movie is, we can't use its principles to turn a real sealed balloon inside out. Because real balls and balloons are not made of a material that can pass through itself, it is not possible to turn such objects inside out without poking a hole through them.

Topologists have also long wondered if it were possible to turn a torus inside out through a hole poked in its side. It turns out that this is a fairly easy operation. We can do it without tearing the torus so long as the torus starts with a hole in its side. Although this surprises some of my friends, because it's difficult to visualize, the eversion has been done with real tire inner tubes. To help study the torus eversion process, we can paint a red ring on the outside of a torus and another on the inside. Looking at figure 5.13 (top), the two rings seem to be interlinked like two rings in a chain. However, during the process of turning the torus inside out, the two rings switch positions without breaking either ring (figure 5.13, bottom).

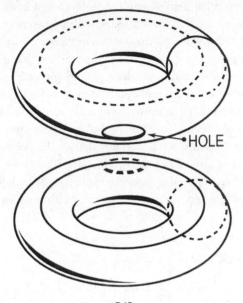

5.13
Turning a torus inside out through a hole in its side.

If you wish to experiment yourself and try to turn a real torus inside out, the task is easier to accomplish with a cloth model than a rubber one. Fold a square cloth in half and sew the opposite ends to make a doughnutlike shape. Now cut a hole in the cloth and push the torus through the hole.

Many perplexing puzzles and transformations exist with tori. For example, if a torus without a puncture hole in its side is linked like a ring in a chain to a torus with a puncture hole (called the "cannibal torus"), the cannibal torus can swallow the torus without the hole so that this torus is completely inside the cannibal torus. In the process of swallowing, the cannibal torus turns inside out. The wonderful torus-swallowing procedure requires the hole in the cannibal to lengthen dramatically. John Stillwell, a mathematician at Monash University, Australia, first showed how this could be achieved by stretching and compressing but, of course, without any tearing.

Beyond the Möbius Strip
The one-sided Möbius strip has many interesting close cousins in the world of topology. For example, a Klein bottle, first described in 1882 by

German mathematician Felix Klein, is an object that has a flexible neck that wraps back *into* itself to form a shape with no inside or outside. This bottle is related to the Möbius strip and can in theory be created by gluing two Möbius strips together along their edges or by gluing both pairs of opposite edges of a rectangle together giving one pair a half twist. In chapter 6, when we consider mathematical models for the entire cosmos, I'll discuss the Klein bottle in more detail and provide illustrations.

Another closely related surface is the real projective plane. It's a closed topological manifold (i.e. surface) that can be visualized by connecting the sides of a square in the orientations illustrated in figure 5.14. In other words, the right edge is twisted relative to the left edge before gluing the edges together, as are the top and bottom edges.

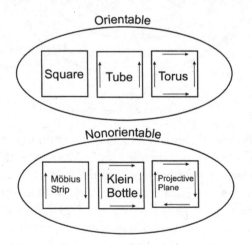

5.14 Schematic diagrams showing how to create various surfaces. In a Möbius strip, we connect two opposite sides of the square with a twist, signified by the arrows in opposite directions. In the real projective plane, pairs of opposite surfaces are connected with a twist.

The real projective plane is a nonorientable surface, which, as with the Möbius strip, means that a creature can travel within the surface and find paths that will reverse its handedness when it returns to its starting point. A Möbius strip is created when we poke a hole in the real projective plane. Other surfaces—with fun names like the "Boy surface," "crosscap," and "Roman surface"—are all homeomorphic to the real projective plane and contain self-intersections when we try to represent these surfaces in our 3-D world.

A cross-cap is a two-dimensional surface that is topologically equivalent to a Möbius strip. To produce such an object, you can imagine sewing the circular edge of a hemisphere to a Möbius strip. The resultant surface has no true inside or outside and looks like a dented, brimless hat. A cross-cap that has been closed up by gluing a disc to its boundary becomes a real projective plane. Two cross-caps glued together at their boundaries form a Klein bottle.

Figure 5.15 shows a way to attach a cross-cap to a surface. First, cut a hole in a surface and sew a Möbius band to it, edge to edge. In three-dimensional space, the only way to do this is to permit the Möbius band to intersect itself. Because a Möbius band changes the handedness of objects that travel within its surface, a cross-cap is also a nonorientable surface.

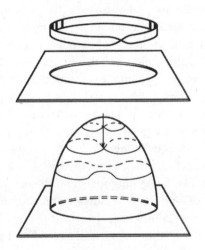

5.15 The cross-cap. To attach a cross-cap to a surface, first cut a hole in the surface and sew a Möbius strip to the surface along its edge.

Other related shapes include Jakob Steiner's "Roman surface" (which looks like a very deformed bowl from one viewpoint) and Werner Boy's "Boy surface" (which, in some orientations, looks like a twisted pretzel) (figure 5.16). The Boy surface, described by Werner Boy in 1901, is a nonorientable surface like the cross-cap and Roman surface, all of which can be obtained by sewing a Möbius strip to the edge of a disk in different ways. Unlike the Roman surface and the cross-cap, it has no singularities (pinch points), but it does intersect itself.

5.16
Boy's surface. (Computer rendition by Jos Leys.)

Möbius Function

August Ferdinand Möbius's interests went far beyond geometry as he explored several exotic integer functions. Throughout the years, I have enjoyed cataloguing these kinds of curious mathematical functions, which have complicated or elegant behavior, and which provide mathematicians fertile territory for future exploration. Most functions that we learn about in high school, like $y = x^2$, which defines a parabola, are rather smooth and exhibit tame behaviors. In this section, let's study a function that has a very irregular behavior and that has intrigued mathematicians since the days of Möbius.

Sometime around 1831, Möbius studied what was later named the Möbius function in his honor. To understand the function, which is represented by the Greek letter mu (μ), imagine placing all the integers into just one of three large mailboxes as described shortly. The first mailbox is painted with a big "0," the second with "+1," and the third with "-1."

0   +1   -1

In mailbox 0, Möbius places multiples of square numbers (other than 1), including {4, 8, 9, 12, 16, 18, 20, 24, 25, 27, 28, 32, 36, 40, 44, 45, 48, 49, 50, 52, 54, 56, 60, 63, 64, etc}. A square number is a number such as 4, 9, 16, or 25 that is the square of another integer. For example, $\mu(12) = 0$ because 12 is a multiple of the square number 4 and is thus placed in mailbox "0."

Before proceeding, I would like to digress because we can already make some remarkable observations. Mathematicians know that the probability that a number is *not* located within the "zero" mailbox tends toward $6/\pi^2 = 0.6079$ . . . as the mailboxes fill up with numbers. Out of the first 100,000 numbers, this $6/\pi^2$ probability predicts 39,207 numbers with $\mu(n) = 0$. The actual figure is 39,206. It always amazes me that π frequently appears in mathematical areas seemingly unrelated to π's original application in geometry.

Let's take another peek in mailbox 0 with its square-containing numbers (also called "squareful" or "nonsquarefree" numbers). Notice that the first occurrence of two consecutive numbers occurs at {8, 9}. Three numbers occur in a row in the previous list at {48, 49, 50}. It's possible to list the smallest term in the first run of at least n consecutive integers that are not squarefree:

- 4
- 8
- 48
- 242
- 844
- 22,020
- 217,070
- 1,092,747
- 8,870,024
- 221,167,422
- 221,167,422
- 47,255,689,915
- 82,462,576,220
- 1,043,460,553,364
- 79,180,770,078,548
- 3,215,226,335,143,218
- 23,742,453,640,900,972

Notice that the terms for $n = 10$ and $n = 11$ are the same, namely 221,167,422. I do not know if mathematicians have ever found two consecutive ns like this anywhere else in the sequence. (An interesting factoid: No squareful Fibonacci numbers F_p are known with p prime.)

Now, let us return our attention to the Möbius function and the mailboxes. The fundamental theorem of arithmetic tells us that every positive integer factors into a unique set of prime numbers. For example, 30 is the product of 2, 3, and 5. In the -1 mailbox, Möbius places any number that factors into an odd number of distinct primes, such as {2, 3, 5, 7, 11, 13, 17, 19, 23, 29, 30, 31, 37, 41, 42, 43, 47, 53, 59, 61, 66, 67, 70}. For example, $5 \times 2 \times 3 = 30$, so 30 is in this list because it has three prime factors. All prime numbers are also on this list because they only have one prime factor, themselves. Thus, $\mu(29) = -1$ and $\mu(30) = -1$.

The probability that a number falls in the -1 mailbox is $3/\pi^2$, which we may write as $P[\mu(n) = -1] = 3/\pi^2$. Here is yet another intriguing occurrence of π far from its traditional geometrical interpretation.

Finally, let's consider the +1 mailbox in which Möbius places all the numbers that factor into an even number of distinct primes. For completeness, Möbius put 1 into this bin. Numbers in this mailbox include {1, 6, 10, 14, 15, 21, 22, 26, 33, 34, 35, 38, 39, 46, 51, 55, 57, 58, 62, 65, 69, 74}. For example, 26 is in this mailbox because $26 = 13 \times 2$. From our discussion, you can see that the Möbius function has a value of 1 or -1 only if no prime is repeated in a number's factorization. The probability that a number falls in the +1 mailbox is $3/\pi^2$.

Given this long introduction, we can list the first twenty terms of the wonderful Möbius function: $\mu(n) = \{1, -1, -1, 0, -1, 1, -1, 0, 0, 1, -1, 0, -1, 1, 1, 0, -1, 0, -1, 0\}$. When we plot this function (figure 5.17), it "looks" random in the sense that it seems to be chaotic with no discernible pattern or regularity.

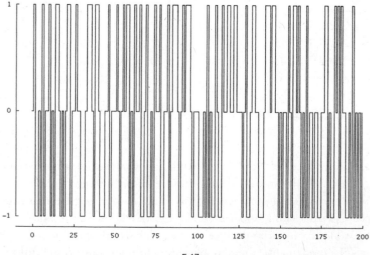

5.17
The erratic Möbius function, $\mu(n)$, for values of n up to 200. (Graph by Mark Nandor.)

The cumulative sum for $\mu(n)$ is {1, 0, -1, -1, -2, -1, -2, -2, -2, -1, -2, -2, -3, -2, -1, -1, -2, -2, -3, -3}, which is known as the Mertens function, or $M(x)$. Figure 5.18 shows the Mertens function for the first 100,000 values.

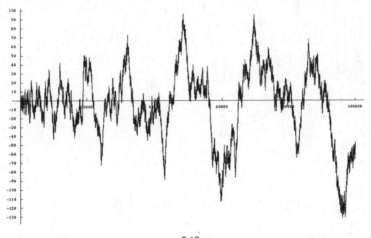

5.18

Mertens function $M(x)$ for values of x up to 100,000. (Graph by Mark Nandor.)

In 1897, European mathematician Franz Mertens made the bold conjecture that $|M(x)/x^{1/2}| < 1$ for all x. In other words, he asserted that the absolute value of $M(x)$ would always be less than the square root of x. Mertens made a table of values for both $\mu(n)$ and $M(n)$ that was fifty pages long and included values for n up to 10,000.

Mertens peered long and hard at the list, and as he compared $M(n)$ to n, he made his famous conjecture. In 1897, mathematician R. D. von Sterneck conjectured that $|M(x)/x^{1/2}| < 1/2$ after he arduously calculated $M(x)$ for x running up to five million and found that $|M(x)/x^{1/2}| < 1/2$ was always true after the first two hundred values. Figure 5.19 shows $M(x)/x^{1/2}$. Notice how the value never goes beyond negative or positive 0.5 after the first few hundred values.

Years later, the Sterneck conjecture was discovered to fail. In particular, for $x > 200$, the first time $|M(x)/x^{1/2}|$ exceeds 1/2 is at $M(7,725,030,629) = 43,947$, discovered in 1960 by Wolfgang Jurkat. In 1979, H. Cohen and F. Dress computed the values of $M(x)$ for x up to 7.8 billion and still the original Mertens conjecture held!

It wasn't until 1983 that Herman te Riele and Andrew Odlyzko disproved the Mertens conjecture that $|M(x)/x^{1/2}| < 1$ for all x. Mertens function expert Ed Pegg Jr. tells me that it wasn't until 1985 that Andrew

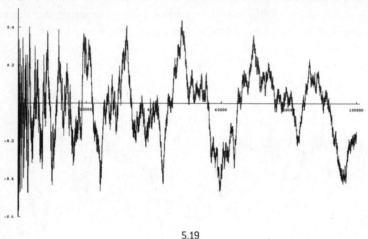

5.19

$M(x)/x^{1/2}$ (1 < x < 10,000). (Graph by Mark Nandor.)

Odlyzko finally found an actual example near $x = 10^{10^{64.1442}}$ where $|M(x)/x^{1/2}| > 1.06$. It is estimated that the first number x that fails the Mertens conjecture is greater than 10^{30}.

In 1987, J. Pintz showed that another Mertens counterexample could be found for some x less than 10^{65}. The *first* value for which $|M(x)/x^{1/2}| > 1$ is still not known. In 1985, Odlyzko and Riele believed that there were no counterexamples to the Mertens conjecture for $x < 10^{20}$.

The Möbius function is fascinating, in part, because of the number of elegant and profound identities that mathematicians have found that involve it. Here are just a few:

$$\sum_{n=1}^{\infty} \frac{\mu(n)}{n} = 0$$

$$\sum_{n=1}^{\infty} \frac{\mu(n)\ln n}{n} = -1$$

$$\sum_{n=1}^{\infty} \frac{|\mu(n)|}{n^2} = \frac{15}{\pi^2}$$

$$\prod_{n=1}^{\infty} (1-x^n)^{\mu(n)/n} = e^{-x} \text{ for } |x| < 1$$

Applications

The Möbius function has applications in various areas of physics. For example, scientists have found practical uses of the Möbius function in various physical interpretations of subatomic particle theory. As physicist Donald Spector discusses in his paper "Supersymmetry and the Möbius Inversion Function," the Möbius function can be interpreted as giving the number of fermions in quantum field theory. A fermion is a particle, such as an electron, proton, or neutron, obeying statistical rules requiring that not more than one fermion may occupy a particular quantum state. The fact that $\mu(n) = 0$ when n is not squarefree is equivalent to the Pauli exclusion principle. Spencer writes to me, "Yes, the Möbius function does provide insight into the structure of particle theory, and it is also fair to say that the applications go in both directions, so that my work shows that particle physics can provide insights into number theory."

Readers interested in applications such as these should consult theoretical physicist Marek Wolf's paper "Applications of Statistical Mechanics in Prime Number Theory." Patrick Billingsley, professor emeritus at the University of Chicago Department of Statistics, has used the Möbius function to generate random walks in his paper "Prime numbers and Brownian Motion."

The Möbius function also has deep connections with the distribution of prime numbers and has a simple relationship with the famous Riemann zeta function ζ, which is of paramount importance in number theory because of its relation to the distribution of prime numbers. (While many of the properties of the zeta function are known, several important fundamental conjectures, the most famous being the Riemann hypothesis, remain unproven.) Consider the famous identity

$$\sum_{n=1}^{\infty} \frac{\mu(n)}{n^s} = \frac{1}{\zeta(s)} = \prod_{p=primes} (1 - \frac{1}{p^s})$$

Here, s is a complex number with real part greater than 1, and the product denoted by the $\prod$ symbol is over all primes. More generally, mathematicians have used the Möbius function as a tool to help solve intricate problems in number theory that involve prime numbers.

Mathematicians find the Möbius function fascinating because almost everything about its behavior is unsolved. We don't even know the Möbius value for most numbers with over three hundred digits.

Applications of Old Math
What other applications might Möbius's strip or his function find someday? Certainly, there are many examples of ancient math finding

obscure applications centuries later, and such math has even been used to describe the very fabric of reality. For example, in 1968, Gabriele Veneziano, a researcher at CERN (a European particle accelerator lab) observed that many properties of the strong nuclear force are perfectly described by the Euler beta function, an obscure formula devised for purely mathematical reasons two hundred years earlier by Leonhard Euler. In 1970, three physicists, Nambu, Nielsen, and Susskind, published their theory on the beta function, the precursor to modern string theory, which says that all the fundamental particles of the universe consist of tiny strings of energy.

Möbius Function Palindromes ("Möbidromes")

My colleague Jason Earls from Fritch, Texas, author of *Death Knocks*, is one of the world's experts on the Möbius function when applied to palindromes, numbers that read the same left to right and right to left like 12,321. One of his pleasing discoveries, made in 2004, involves the Möbius function applied to the palindrome 15,891,919,851 and each right truncation of its digits.

$$\mu(15{,}891{,}919{,}851) = 1$$
$$\mu(1{,}589{,}191{,}985) = 1$$
$$\mu(158{,}919{,}198) = 1$$
$$\mu(15{,}891{,}919) = 1$$
$$\mu(1{,}589{,}191) = 1$$
$$\mu(158{,}919) = 1$$
$$\mu(15{,}891) = 1$$
$$\mu(1{,}589) = 1$$
$$\mu(158) = 1$$
$$\mu(15) = 1$$
$$\mu(1) = 1$$

He also discovered the following sequence when the Möbius function is applied to the palindrome 79,737,873,797 and each right truncation of its digits:

$$\mu(79{,}737{,}873{,}797) = -1$$
$$\mu(7{,}973{,}787{,}379) = -1$$
$$\mu(797{,}378{,}737) = -1$$
$$\mu(79{,}737{,}873) = -1$$

$$\mu(7{,}973{,}787) = -1$$
$$\mu(797{,}378) = -1$$
$$\mu(79{,}737) = -1$$
$$\mu(7{,}973) = -1$$
$$\mu(797) = -1$$
$$\mu(79) = -1$$
$$\mu(7) = -1$$

Jason spends much of his leisure time searching for Möbius palindromes (or "Möbidromes"), like an astronomer scanning the sky for signs of extraterrestrial life. He does this for no reason that I can ascertain, except for the sheer joy he feels when making discoveries that no one else has ever made. Will he ever find a larger Möbius palindrome? Do infinitely many Möbius palindromes exist (i.e. palindromes that return 1 or -1 for the Möbius function for each right truncation of their digits)?

The Amazing Ubiquity of π

We discussed the remarkable occurrence of π when dealing with the Möbius function, and I'm generally fascinated by the ubiquity of π in far-flung areas of mathematics. Normally we think of π simply as the ratio of the circumference of a circle to its diameter. So did pre-seventeenth-century humanity. However, in the seventeenth century, π was freed from the circle. Many curves were invented and studied–for example, various arches, hypocycloids, and witches–and mathematicians found that their areas could be expressed in terms of π.

Finally, π ruptured the confines of geometry altogether. Today π relates to many areas in number theory, probability, complex numbers, and simple fractions. such as $\pi/4 = 1 - 1/3 + 1/5 - 1/7. \ldots$ It is sometimes difficult to account for its wide sphere of influence.

As an example of how far π has drifted from its simple geometrical interpretation involving circles, consider the book *Budget of Paradoxes*, where Augustus De Morgan explains an equation to an insurance salesman. The formula, which calculates the chances that a particular group of people would be alive after a certain number of days, involves the number π. The insurance salesman interrupts and exclaims, "My dear friend, that must be a delusion. What can a circle have to do with the number of people alive at the end of a given time?"

Satellite photos of rivers yield π in a strange way. Imagine you are examining a photo of the full length of a meandering river. Measure the distance of the river along a straight line connecting the start and end of the river, and call this distance D. Next, measure the distance of the river along its actual length, as if you were traveling by boat. Call this distance R. According to Hans-Henrik Stĩlum, an earth scientist at Cambridge University, π is the average ratio of R to D for meandering rivers. Although the ratio varies from river to river, the average value of $R/D =$ π is most commonly found for rivers flowing across very gently sloping planes, such as found in Brazil or the Siberian tundra. As Simon Singh, author of *Fermat's Last Theorem*, wrote: "In the case of the river ratio, the appearance of π is the result of a battle between order and chaos."

Even more recently, π has turned up in equations that describe subatomic particles, light, and other quantities that have no obvious connection to circles. We have already discussed that the probability that a randomly chosen integer is squarefree (not divisible by a square) is $6/\pi^2$. The value $\pi^2/6$, denoted by λ, is everywhere in mathematics. For example, it appears in the sum of the reciprocals of the squares of the positive integers:

$$\lambda = \frac{\pi^2}{6} = \sum_{n=1}^{\infty} \frac{1}{n^2}$$

The hypervolume of a four-dimensional hypersphere is $3\lambda r^4$. The integral from 0 to infinity of $x/(e^x - 1)dx$ is λ. We also have: The expression

$$\frac{\pi^2}{6} = -2e^3 \sum_{n=1}^{\infty} \frac{1}{n^2} \cos\left(\frac{9}{n\pi + \sqrt{n^2\pi^2 - 9}}\right)$$

$$\frac{\pi^2}{6} = \pi - 1 + \sum_{n=1}^{\infty} \frac{\cos(2n)}{n^2}$$

$6/\pi^2 = 0.608 \ldots$ or its reciprocals shows up in countless seemingly unrelated areas of mathematics, giving it an almost mystical significance. For example, consider that $6/\pi^2$ is also the probability that two numbers selected at random are coprime. (Number theorists call two numbers A and B that have no common factors "relatively prime" or "coprime.") As an example of coprimality, two integers are said to be coprime if their greatest common divisor equals 1. For example, 5 and 9 are coprime, while 6 and 9 are not comprime because their greatest common divisor is 3.

In fact, Clive Tooth is so excited about the fantastic occurrences of $\pi^2/6$ in mathematics and beyond that he has devoted a Web page to this topic: http://www.pisquaredoversix.force9.co.uk/.

Finally, while on the subject of coprimality, I cannot resist the urge to tell you another quick bit of mathematical trivia. A standard function in number theory is $\phi(n)$, which is the number of integers smaller than n and relatively prime to n. Amazingly, we find that:

$$\phi(666)=6\bullet6\bullet6.$$

This should appeal to people looking for odd occurrences of 666, the "number of the beast" in the book of Revelation.

Möbius Strip and Graph Theory

Draw several dots on a piece of paper. What is the largest number of dots that can be joined by lines that do not intersect and that connect every pair of points? (The paths you draw to connect the points may curve.) With just two points, we can connect "all the points" with one line (figure 5.20). With three points, we can connect all the pairs of points to form a triangle. With four points, we can still manage to connect every possible pair of points. Just how far can we go?

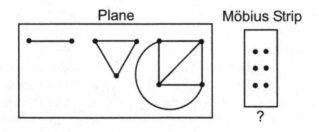

Plane   Möbius Strip

5.20
On a plane and on a Möbius strip, what is the largest number of dots that can be joined by lines that do not intersect and that connect every pair of points?

It turns out that four is the largest number of dots, and we can't succeed in connecting all pairs with five dots drawn on a plane. However, the situation gets more interesting if we ask the identical question for dots on a Möbius strip. Can you solve this before reading further? Can you connect the six dots in figure 5.20 on a Möbius strip with lines that do not intersect and that connect every pair of points? When we talk about dots on a Möbius surface, we must think of the surface as having no thickness so that each line is embedded in the surface like a magic marker line that penetrates the paper all the way through.

Figure 5.21 is one solution to the six-dot graph problem on a Möbius strip and is discussed in Martin Gardner's *Mathematical Magic Show*. To understand how the diagram illustrates the connectivity of six dots, assume that the right and left sides of the strip are connected after a half twist. Again, the surface is thought of as having zero thickness with lines "in" it in the same way that humans are in their 3-D space. Are there other elegant, symmetrical solutions to this problem?

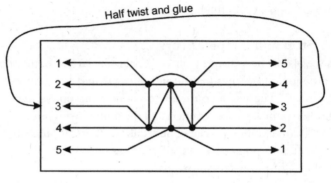

5.21
One symmetrical solution to the six-dot graph problem on a Möbius strip.

Hexaflexagons

Hexaflexagons are geometrical objects that have an odd number of half twists and are therefore Möbius surfaces. Martin Gardner made hexaflexagons famous in *Hexaflexagons and Other Mathematical Diversions: The First Scientific American Book of Puzzles and Games*. In the book he described these elegant paper hexagons that fold from strips of paper and reveal different faces as they are flexed. They were first discovered in 1939 by Arthur Stone, who set up the Flexagon Committee, which brought together famous mathematicians and physicists to investigate the properties of these unique shapes. You can learn more about the amusing forms using the Google Web search engine.

Other One-sided Surfaces

Examples abound for one-sided surfaces with just one edge ("a" and "b" in the top row of figure 5.22) and two edges (the remaining six figures). The surfaces may be knotted or unknotted and edges may be linked or unlinked. The top left figure (a) is a Möbius strip.

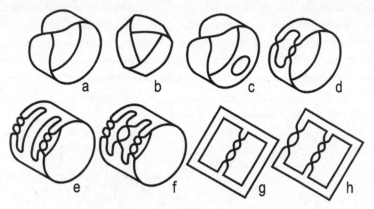

5.22

A zoo of one-sided surfaces. Top row: a) edge is a simple closed curve; b) edge is knotted; c) both edges are simple closed curves, unlinked; d) both edges are simple closed curves, linked; e) both edges are knotted, unlinked; f) both edges are knotted, linked; g) one edge is simple, one knotted and unlinked; h) one edge is simple, one knotted and linked. (After David Wells, *The Penguin Dictionary of Curious and Interesting Geometry*.)

For these shapes, you can understand what it means to have a "knotted edge" by visualizing the edge as a piece of string. If the knotted edge were made of string, it couldn't be untangled to form a simple circular loop without cutting. If the edges are "linked," then the edge consists of more than one piece of string linked so they can't be separated without cutting. More generally, a curve is knotted if it cannot be deformed into a circle without cutting it. Two curves are linked if they cannot be separated without cutting one of them.

For the Möbius strip, if the central "paper part" disappeared and the edge of the strip is visualized as a string, the string could be stretched into a circle. However, in the case of a strip with three half twists, if the surface disappears and the edge is turned into a piece of string, the string is tangled.

Möbius Shorts

Möbius shorts are one-sided surfaces reminiscent of the Möbius strip. I'm not sure who first contemplated the Möbius shorts shown in figure 5.23, but several sources attribute it to an unknown researcher named Gourmalin. I came across this wonderful object while reading Ralph Boas Jr.'s article titled "Möbius Shorts" published posthumously in a 1992 *Mathematics Magazine*. Boas says that he discovered this in the *Dictionnaire des mathématiques* by Alain Bouvier, Michel George, and François Le Lionnais (Paris, 1979). This surface is topologically equivalent to a Klein bottle with a hole in it and is topologically distinct from the Möbius strip.

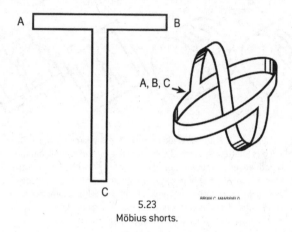

5.23
Möbius shorts.

If you want to construct a paper model, start with the T-shaped piece of paper shown in figure 5.23. Bend the top of the T to make an untwisted ring, and glue A to B. Pass C upward through the ring, turn C down (without twisting), and glue C to the outside of the ring at AB. The result is a one-sided surface. Try coloring it. What happens if we cut both the ring and what was originally the stem along their midlines? Boas claimed that neither the Möbius shorts nor the results of cutting are well-known in American mathematical circles.

Möbius Tetrahedra

A regular tetrahedron looks like a pyramid with a triangular base. The object has four vertices, six edges, and four equivalent equilateral triangular faces. Möbius explored a class of tetrahedra, now called Möbius tetrahedra in his honor. In particular, Möbius tetrahedra are a pair of tetrahedra, each of which has all its vertices lying on the faces of the other. These tetrahedra are not "regular" with identical facets, but each tetrahedron is inscribed in the other. (In mathematics, "inscribing" usually refers to drawing one figure within another figure so that every vertex of the enclosed figure touches the outer figure.) Möbius discusses these tetrahedra in his 1828 paper *"Kann von zwei dreiseitigen Pyramiden eine jede in Bezug auf die andere um- und eingeschrieben zugleich heissen?"* (rough translation: "Can two three-sided pyramids that inscribe one another be called identical?" or "If two three-sided pyramids can be rotated and translated into one another, can they be called identical?"), and he shows how the strange geometric situation for Möbius tetrahedra can be realized when some of the vertices lie in the extensions of the facial planes. The precise arrangement of mutually inscribing Möbius tetrahedra is extremely difficult to visualize, and readers are urged to test their powers of visualization by studying the "Möbius Tetrahedra" entry at http://mathworld.wolfram.com.

Möbius Triangles

Möbius triangles are triangles on the surface of a sphere. These spherical triangles result when a sphere is divided by the planes of symmetry of a uniform polyhedron. Figure 5.24 shows an example.

5.24 Möbius triangles.

This object has 120 Möbius triangles. Each triangle corresponds to one tenth of a dodecahedron face or, equivalently, one sixth of an icosahedron face. Black and white indicates left- and right-handed triangles. In other words, the black and white triangles are mirror images of each other, also known as enantiomorphs. You can learn more about Möbius triangles at George Hart's "Millennium Bookball" Web page, which contains photos of his sculptures that are reminiscent of Möbius triangles.

The Solenoid

The Möbius strip becomes a springboard to other mathematical adventures. After years of studying the Möbius Strip, I became interested in strange and beautiful computer graphics generated when studying other twisted topological forms. One of my favorite shapes is the solenoid, a weird, twisted doughnutlike shape. It's a topological construction that arises from, and is related to, a famous fractal called the Cantor set. It is also one of the principal examples of a "strange attractor" in dynamical systems theory. In this section, we won't dwell on its interesting topological properties, which would take many pages (see references for further reading). Instead, we can develop some formulas that help elucidate its self-similar structure and facilitate the computer graphical generation of images that are pleasing in their simplicity and grace, yet sufficiently complex to intrigue the eye.

The starting point of the solenoid is the solid torus, followed by a strange transformation of the torus. Here's the best way to visualize this. The mapping squeezes the tube of the torus to half its original diameter, stretches it out to twice its original length, and wraps this length twice around, inside the skin of the original torus. In wrapping around twice, one coil sits next to another one with no overlap, just as one would coil up lengths of a garden hose. The coil makes a half twist as it wraps around once, joining back up to itself after two turns.

I explored the solenoid's form with mathematician Kevin McCarty. We found that the representation of nested tori provides quite a visualization challenge. Some of our graphics showed the solenoids with varying degrees of twisting inside the transparent shell of the standard torus in which it resides, like an embryonic snake squeezed within a toroidal egg. Figure 5.25, shows an example of a solenoid with the toroidal shell removed for clarity. I rotate this on my computer screen so I can observe it from all angles.

5.25
The solenoid.

These strange objects can be continually twisted. Like a taffy machine with no off switch, the operation of stretching, winding and twisting can be repeated indefinitely. As the mapping carries the original torus to an image of itself wrapped twice around, it also carries the twice-wrapped image to one wrapped four times around. Each iteration produces another tube nested inside the previous one. At each stage, the number of windings doubles and the thickness halves. This process converges in the limit to a connected set of infinitely thin windings, the "final" solenoid.

The easiest way to describe the way this mapping works is to use complex numbers with a real and imaginary component. If this makes little sense to you, turn to the reference section where I have included an outline showing how a computer recipe would work. A point inside the solid torus is located by a pair of complex numbers (z, w). The z coordinate represents the longitude angle and locates a point on the unit circle in the complex plane that will be the center or spine of the torus. The w coordinate locates a point inside a disk of radius 1/2, considered as a piece of the complex plane. The disks are imagined to be threaded on the unit circle like a necklace. With these coordinates, the mapping that wraps the torus twice around inside itself is

$$f(z, w) \longrightarrow (z^2, w/2 + z/4)$$

The term z^2 simply wraps the unit circle twice around itself as z traverses the unit circle once. The term $w/2$ shrinks the original w coordinate to half its size, while the $z/4$ term moves it away from the $w = 0$ origin so the image does not intersect itself on the second loop. The simple algebraic formula allowed by complex number representation makes it easy to compute repeated iterations of the mapping as shown in the iterative program in the reference section.

If we were to take a cross section of the solenoid construction perpendicular to the windings, we would see a sequence of nested disks; each disk contains two smaller disks. When the longitude angle is zero $(z = 1 + 0i)$, all nested disks line up. But for other longitude angles, the varying amounts of twist cause the disks to become separated. This separation can be seen in figure 5.25, which shows the mapping iterated to the second level of nesting.

We can also represent the creation of increasingly intricate solenoids using nomenclature common in the topological literature. Consider the map on the solid torus given by

$$F_\gamma(\theta, x) = (2\theta, \gamma x + \tfrac{1}{2} e^{i\theta})$$

We can visualize what this map means by imagining cutting a torus with a sharp knife once to create a long cylinder. Next, we stretch the cylinder to twice its length while contracting its width by γ. Wrap the resulting long, thin cylinder around itself twice, rejoin the sticky ends, and replace it inside the original torus space. Iterating the solenoid map n times results in a spindly tube that winds around the inside of the original "fat" torus 2^n times. For additional background on the solenoid, consult Stephen Smale's "Differentiable Dynamical Systems," which describes his identification of this kind of object as an example of a strange attractor.

The Horned Sphere

As we've discussed, the Möbius strip is an example of an object with one surface, and the Klein bottle is an object with no distinct inside or outside. In addition to these shapes, mathematicians continue to invent strange objects to test their intuitions. Alexander's horned sphere is an example of a convoluted, intertwined surface for which it is difficult to define an inside and outside. Introduced by mathematician James Waddell Alexander (1888–1971), Alexander's horned sphere (figures 5.26–5.28) is formed by successively growing pairs of horns that are almost interlocked and whose end points approach each other. The initial steps of the construction can be visualized with your fingers. Move the thumb and forefinger of each of your hands close to one another, then grow a smaller thumb and forefinger on each of these, and continue this budding without limit!

Although this may be hard to visualize, Alexander's horned sphere is homeomorphic to a ball. In this case, this means that it can be stretched into a ball without puncturing or breaking it. Perhaps it is easier to visualize the reverse: stretching the ball into the horned sphere without ripping it. The boundary is, therefore, homeomorphic to a sphere.

5.26
Alexander's horned sphere. (Image created by Cameron Browne.)

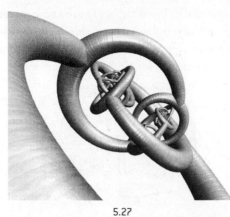

5.27
Magnification of Alexander's horned sphere. (Image created by Cameron Browne.)

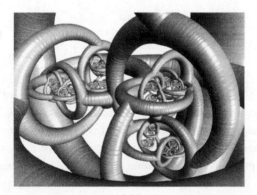

5.28
Magnification of Alexander's horned sphere. (Image created by Cameron Browne.)

Figure 5.29 is Cameron Browne's representation of Alexander's horned sphere embedded in the plane. This "woven horn" is based on Alexander's horned sphere, which, as we discussed, is traditionally visualized as a recursive set of interlocking pairs of orthogonal rings of decreasing radius. Cameron embeds his woven horned sphere in the plane by reducing the interlock angle between ring pairs from 90 degrees to 0 degrees. Next, he creates an over-under weaving pattern to reestablish the ring interlock without intersection to produce the woven horn set. Browne tells me that the woven horn construction is a self-similar fractal but not technically an area-filling curve, because any open subset of the plane will contain points that are a nonzero distance from the curve.

Cameron Browne is a professional software engineer with degrees in computer science and psychology, two fields he hopes to unite in his future research. He has spent the last few years researching automatic font decoration and animation for Canon and Microsoft.

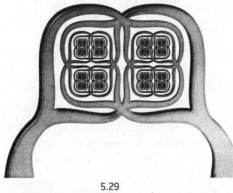

5.29
Cameron Browne's representation of Alexander's horned sphere embedded in the plane.

On the Wonders of Prismatic Doughnuts

While writing this book, I became obsessed with the concept of prismatic doughnuts, and I experimented with the object in figure 5.30, which I created in clay. I invented this strange object, but owing to its simplicity I suspect that mathematicians have investigated it numerous times. Consider a hexagonal prism, with two hexagonal ends, that normally has six distinct faces, not counting the hexagons at each end. Now consider a twisted hexagonal prism made out of clay, whose two hexagonal end plates are joined

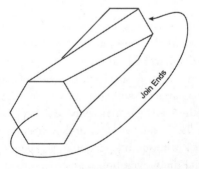

Join Ends

5.30
How to create a hexagonal, twisted prismatic doughnut with only one face.

together like a doughnut. I asked colleagues to ponder this question: "How many faces does this new doughnutlike structure have, given the twist shown in the diagram?" Is it possible to make a hexagonal doughnut with just one unique face if the prism's back end plate is given the proper number of twists relative to the front end plate and then joined to form the doughnut?

It turns out that the hexagonal prismatic doughnut specified in figure 5.30 has just one face! This assumes that there is no additional twisting of the end plates during the joining process. For example, you can start traveling down one "road," say face 1, and end back on face 1 after touching each apparently different face exactly once.

In figure 5.30, if we number the sides consecutively and rotate one end by sixty degrees, then side 1 will connect to side 2, side 2 to side 3, side 3 to side 4, side 4 to side 5, side 5 to side 6, and side 6 to side 1. We will be able to reach any point on the surface from any other without crossing an edge, so the object is a one-sided surface. This image has a sixty-degree clockwise twist of the back plate relative to the front plate. Similarly, twisting the back plate by the same counterclockwise increment will also create a "Möbius prismatic doughnut." In addition, twisting 5/6 of the way around in either direction will yield a single face, as will 7/6, 11/6, and so forth.

When I showed the Möbius prism to my colleague Mark Nandor, he conjectured that, for a prism with n faces, there are $\phi(n)$ different twists (clockwise or counterclockwise) that will yield a single-face, plus any twists that are the same fraction rotation beyond a full twist. Euler's phi, denoted $\phi(n)$, is the number of numbers less than n that are relatively prime to n. As discussed, number theorists call two numbers A and B that have no common factors (except 1) "relatively prime" or "coprime." For instance, $\phi(6) = 2$, since 1 and 5 are relatively prime to 6, and $\phi(10) = 4$ since 1, 3, 7, and 9 are relatively prime to 10. So, for a six-sided prismatic doughnut, there are two twists in the clockwise direction (1/6 of a twist and 5/6 of a twist) that will yield one face. Note that not only can we use 7/6 and 11/6 twists, but we can also twist in the opposite direction.

In summary, for a prism with n faces, there are $\phi(n)$ different fractions of twists in the clockwise direction that will yield a single face once the ends are joined. Each fractional twist corresponds to k/n, where k is any number smaller than n that is also relatively prime to n (the greatest common factor of k and n is 1). Note that the prism could also be twisted in the counterclockwise direction as well. Lastly, the prism could be twisted more than once around, so a $1 + k/n$ twist and a $2 + k/n$ twist (or any $N + k/n$ twist) will also yield single-sided shapes.

For a prism with n faces, there are $\phi(n)$ modulo n different fractions of twists in either direction that will yield a single-sided figure once the end-plates are joined.

$\phi(n)$ has an interesting formula. If the prime factorization of n is $n = A^a \times B^b \times C^c \times \ldots$ then

$$\phi(n) = n[(A\text{-}1)/A][(B\text{-}1)/B][(C\text{-}1)/C] \ldots$$

So, for $6 = 2 \times 3$, $\phi(6) = 6 \times (1/2) \times (2/3) = 2$. As we have seen, the two relatively prime numbers to 6 that are less than 6 are 1 and 5. For $300 = 2^2 \times 3^1 \times 5^2$, $\phi(300) = 300 \times (1/2) \times (2/3) \times (4/5) = 80$. The numbers less than 300 relatively prime to 300 are 1, 3, 7, 9, 11, 13, 17, 19, 21, 23, 27, 29, 31, 33, 37, 39, 41, 43, 47, 49, 51, 53, 57, 59, 61, 63, 67, 69, 71, 73, 77, 79, 81, 83, 87, 89, 91, 93, 97, 99, 101, 103, 107, 109, 111, 113, 117, 119, 121, 123, 127, 129, 131, 133, 137, 139, 141, 143, 147, 149, 151, 153, 157, 159, 161, 163, 167, 169, 171, 173, 177, 179, 181, 183, 187, 189, 191, 193, 197, 199, 201, 203, 207, 209, 211, 213, 217, 219, 221, 223, 227, 229, 231, 233, 237, 239, 241, 243, 247, 249, 251, 253, 257, 259, 261, 263, 267, 269, 271, 273, 277, 279, 281, 283, 287, 289, 291, 293, 297, and 299.

As a final effort in visualization, look at the hexagonal twisted prism in figure 5.30, and imagine gluing to each face a triangular prism so that each face of the hexagonal prism is coplanar, with one face of each triangular prism. This would mean that each face in the hexagonal prism is now split into two new faces by gluing the triangular prism to the face. What properties does this hyperprismatic doughnut exhibit when its ends are glued together?

Perfect Square Dissection of a Möbius Strip

A difficult puzzle that has captivated mathematicians for at least a hundred years involves the operation of "squaring a square," also known as a "perfect square dissection." The general problem is to tile a square using square tiles all of *different* sizes. This may sound easy, and you can even experiment with a pencil, paper, and graph paper, but it turns out that very few tile arrangements work.

The first squared *rectangle* was discovered in 1909 by Z. Morón. Morón found a 33 by 32 rectangle, which uses nine squared tiles of sides 1, 4, 7, 8, 9, 10, 14, 15, and 18. He also discovered a 65 by 47 rectangle tiled with ten square tiles of 3, 5, 6, 11, 17, 19, 22, 23, 24, and 25 (figure 5.31). For years, mathematicians claimed that perfect square dissections of squares were impossible to construct.

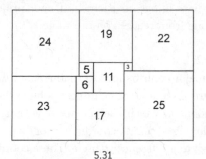

5.31
Morón tiling of a rectangle with squares each of different sizes.

In 1936 four students at Trinity College–R. L. Brooks, C. A. B. Smith, A. H. Stone, and W. T. Tutte–became fascinated by the topic, and finally, in 1940, these mathematicians discovered the first squared square consisting of sixty-nine tiles! With further effort, Brooks reduced the number of tiles to thirty-nine. In 1962, A. W. J. Duivestijn proved that any squared square must contain at least twenty-one tiles, and in 1978 he had found such a square and proved that it was the only one.

In 1993, S. J. Chapman found a tiling of the Möbius band using just five square tiles whose boundaries do not meet themselves when the band's edges are glued (figure 5.32). The arrows in the figure show the two edges of the strip that are glued together with a twist. A cylinder can also be tiled with squares of different sizes, but this requires at least nine tiles.

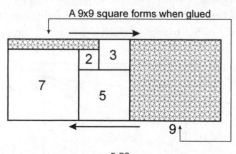

5.32
Different-sized square tiles on a Möbius band.

When we attempt to tile a cylinder or Möbius strip, we must do so with the tile edges parallel or perpendicular to the edges of the surfaces. However, the torus, Klein bottle, and projective plane don't have edges, so the tiles can be inlaid at any angle. I have not been able to find much information on square tilings on Klein bottles or projective planes and would enjoy hearing from readers on the subject.

Barycentric Calculus

One of Möbius's major contributions to mathematics was his barycentric calculus, a geometrical method for defining a point as the center of gravity of certain other points to which coefficients or weights are ascribed. We can think of Möbius's barycentric coordinates (or "barycentrics") as coordinates with respect to a reference triangle. These coordinates are usually written as triples of numbers corresponding to masses placed at the vertices of the triangle. In this way, these masses determine a point, which is the geometric centroid of the three masses. The new algebraic tools developed by Möbius in his 1827 book *Der Barycentrische Calcul (The Barycentric Calculus)* have since turned out to have wide application.

I'll try to make this concept clear with an illustration. The word barycentric is derived from the Greek *barys* (heavy), and refers to the center of gravity. Möbius understood that several weights positioned along a straight stick can be replaced by a single weight at the stick's center of gravity. From this simple principle, he constructed a mathematics system in which numerical coefficients are assigned to every point in space.

In the process of developing his barycentric coordinates, Möbius visualized points with weights as in figure 5.33. Imagine a line *AB* on a plane. Let us first dangle weights only at *A* and *B*. The center of gravity lies somewhere between *A* and *B* along the line that joins them. Next we dangle a weight at *C*, and the center of gravity *P* will be pulled away from line *AB* toward the middle of the triangle *ABC*. In particular, the center of gravity

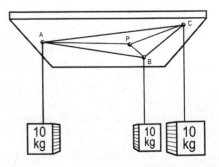

5.33
Barycentric coordinates. Point *P* is the barycenter of *a*, *b*, and *c*, and we say that the "barycentric coordinates" of *P* are (*a*, *b*, *c*).

is pulled in the direction of *PC*. In fact, the triangle will balance on a thin razor blade placed along the line *PC*. It also balances along a razor aligned on lines *PA* or *PB*. The center of gravity *P* is called the barycenter, and triangle *ABC* balances on a pin placed beneath the barycenter.

Möbius then went into great mathematical depth in *Der Barycentrische Calcul* to show several advantages of the use of barycentric coordinates. We can restate his principles by noting that for any point *P* inside a triangle *ABC*, there exist three masses w_A, w_B, and w_C. If placed at the corresponding vertices of the triangle, the center of gravity (barycenter) of these three masses will coincide with the point P. Möbius considered w_A, w_B, and w_C as the barycentric coordinates of P. As defined, the barycentric coordinates are not unique. Masses kw_A, kw_B, and kw_C have exactly the same barycenter for any $k > 0$.

A full accounting of the *usefulness* of barcentric calculus in mathematical theory is beyond the scope of this book, and the reader is urged to see Jeremy Gray's "Möbius's Geometrical Mechanics" for an accessible introduction. It turns out that barycentric coordinates are a form of general coordinates that are used in many branches of mathematics and even computer graphics. If we add one additional constraint, namely, $w_A + w_B + w_C = 1$, then the barycentric coordinates are defined uniquely for every point inside the triangle. Many of the advantages of barycentric coordinates occur in the field of projective geometry, which is concerned with "incidences," that is, where elements such as lines, planes, and points do or don't coincide. Projective geometry is also concerned with the relationships between objects and the mappings that result from projecting them onto another surface. As a visual metaphor, consider that shadows are the projections of solid objects. Barycentric coordinates also arise naturally whenever variable quantities have a constant sum.

In his article "Barycentric Calculus," Alexander Bogomolny, former associate professor of mathematics at the University of Iowa, gives a number of practical examples dealing with probability and puzzles. In particular, he discusses the problem in which we are given three glasses, *A*, *B*, and *C*, of respective capacities 8, 5, and 3 ounces. The first glass is full of water. The problem is to measure out 4 ounces of water. His solutions involve barycentric coordinates by visualizing the points *A*, *B*, *C* at the vertices of a triangular grid. *A*, *B*, and *C* are associated with barycentric coordinates *u*, *v*, *w*, such that $u + v + w = 8$. Bogomolny then uses three-digit strings that correspond to the coordinate values. For example, the apex *A* is referred to by its coordinate string "800," which is just a shorthand for $u =$

8, $v = 0$, $w = 0$, or $(8,0,0)$. Pouring from one glass to another corresponds to moving from one node to another along one of the triangle's grid lines. His internationally acclaimed math Web site www.cut-the-knot.org gives all the mathematical details. Suffice it to say that Möbius's barycentric coordinates have influenced several areas of theoretical and applied mathematics.

◉ Squiggle Map Coloring Puzzle

Nina has created a new form of life with a peculiar kind of skin. She calls the lizard-like animals "morphs" because she can actually design their skin pattern simply by drawing with a felt-tipped pen on their backs. The morphs absorbs the dye pattern, and all their offspring will have the same design. The colorful creatures are becoming all the rage with schoolchildren. Scientists are wondering how it is possible for the morphs' offspring to be born with the same skin design as their parents. Today, Nina draws a maplike squiggle on a morph using a continuous line, not taking her marker off the skin until she returns to her starting point. Figure 5.34 shows an example of Nina's latest design—just one of many she will produce in the coming months. Now it's time for her to color the design.

If Nina is trying to make sure that no contiguous regions are colored the same, what is the minimum number of colors she will need? (In her coloring, two adjacent regions can share a common vertex and have the same color, but they can't share the same edge and have the same color.) Turn to the solutions section for an answer.

5.34
Squiggle coloring. What is the fewest number of colors she needs to produce a design such that any regions with a common boundary line have different colors?

◉ Pyramid Puzzle

In this chapter, we've discussed faceted objects, like tetrahedra, Möbius's house, and twisted prismatic doughnuts. This puzzle tests your powers of visualization. Jill has decorated her bedroom with a huge, colorful pyramid, which has four faces that are equilateral triangles. Jill has painted each face a different color, either red, purple, green, or yellow.

Jill has a brainteaser for you. As she rotates the pyramid, five different views of the pyramid's four corners can be seen from above. Which of the views in figure 5.35 is incorrect? (Turn to the solutions section for the answer.)

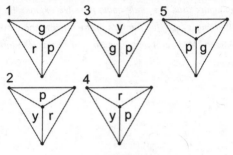

5.35
Several views of a triangular pyramid. Which view is incorrect?

Möbius in Pop Culture

🖉 *Pretty soon, you will never be more than a three-minute drive from a place where you can purchase the following products: a mocha Frappuccino, a chicken burrito the size of your head, NASCAR memorabilia, a cell phone, or an oil change. The entire universe will be one Möbius-strip mall without beginning or end.*

—Mark Hasty, the Bemusement Park Blog

🖉 *Described as an intergalactic take on Jack and the Beanstalk,* Through the Möbius Strip *is the story of physicist Simon Weir, who becomes lost in a space-time portal he created. His son, Jac Weir, must search for him through a myriad number of planets, filled with wondrous sights and often gigantic beings.*

—Animated-news.com

✸ *The pulp, coke, dado, and inert tableting aid are then combined into a solid cake, blue-grey in color, which passes through the immersion font and between a series of pinch rollers and thousands of tiny idler wheels, to emerge on an endless belt, twisted into a three-sided Möbius strip for equalization of wear, where workers toil day and night at adding the curlicued frosting accents that make every snack a special treat*

—Matthew McIrvin, McIrvin's Push-Button World of the Future

✸ *The tale's action ends virtually in the same place it started—Henry standing at his bedroom window, staring at a dark London sky dotted with airplanes—seemingly coming full circle, but like a Möbius strip, this circle has its twists and the route seems longer than the circle could possibly explain. He stands at the open window, shivering, seeing his family's future cut into the predawn sky.*

—Randy Michael Signor, "One Day in February: Metaphor for a Life,"
Chicago Sun-Times

CHAPTER 6

COSMOS, REALITY, TRANSCENDENCE

By considering the mere possibility of higher spatial dimensions, Kant and Möbius each raised interesting philosophical questions. Does the diversity of the cosmos allow for worlds of varied dimensionality in regions separate from ours? Does the presence of left- and right-handed objects imply a four-dimensional transformation that converts one to the other?
　　　　　　　　　　　　—*Paul Halpern,* The Great Beyond

It is my intention, through the methods and theorems of geometry herein laid down, to contribute in some measure to the simplification of its investigations and the broadening of its horizons.
　　　　　　　　　　　　—*August Möbius,* The Barycentric Calculus, *1827*

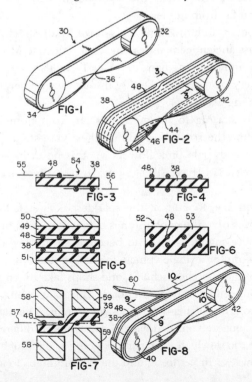

A Journey Beyond New Devonshire

Imagine a land called New Devonshire resembling a typical New England suburb. You've lived in New Devonshire for many years. It's a tranquil place, and you enjoy wandering its tree-lined streets—past ancient churches and rustic parks. New Devonshire is only an hour north of New York City, but it might as well be in an entirely different universe. The town has little pollution, no skyscrapers, few honking horns. The pride and joy of New Devonshire is its bucolic Main Street, with the elegant Möbius Memorial Library, cottage-style homes, and landscaping with meandering pathways, benches, water fountains, and narrow ivy-covered passageways between quaint stores. You've always been happy here.

One day, perhaps because you seek adventure, you decide to ride a bike along Main Street farther then you have ever ridden before, farther, in fact, than anyone in New Devonshire has ever ridden before.

"Good luck," says Linda, the town supervisor. She and a hundred other friends wave good-bye as you speed away down the street.

You're a little nervous. Like most people in New Devonshire, you haven't traveled far from home.

The wind begins to blow along leaf-strewn Main Street. You shiver a little as you pass ancient cemeteries, the Victorian-style Möbius Library, and columned homes dating back to the early 1900s.

Blackbirds cry overhead as they fly through vague perpetual clouds. You smell burning leaves.

The road is starting to fragment this far from home. Tall *Lagerstroemia* trees line the road. You admire their vase-shaped trunks that sport branches with pale pink flowers. At this point in your journey, none of the crossroads have street signs. In fact, there are no signs of any kind.

"Pretty," you say, looking at the sparkles of light falling from the shops and houses onto the cobblestones below your bike tires.

In a few more minutes, the road becomes so narrow at places that there is only room for a single vehicle to pass on it at a time.

The walls of the churches, schools, and shops take on an almond hue. You feel a tingling sensation along your back, a pleasant feeling. This church. This school yard. The butcher selling sausages. A nun. A young man kissing a woman. You squeeze your handlebars. Something is different. Children play, but a little too slowly, hopping and skipping as if through molasses.

The air grows warm yet fresh, and the sky brightens to a pearlescent azure. You look up. The light from the upper stories of the buildings make the structures look inviting, cozy.

You slow your bike, hearing the sound of a church bell; three clangs, then a pause; one clang, then a pause.

Ten minutes later, only small houses line the path. The simple dwellings are a hodgepodge of one-story buildings surrounded by wooden fences. Occasionally, children giggle and point at you. A donkey carrying sacks and a few pieces of wood passes you on its way downhill, almost jostling you into a muddy puddle.

Where is this strange adventure headed? Your fingers tremble. Perhaps this was not a good idea. But then you notice that the road is returning to its modern state. Normal looking street signs appear at the intersections. A car whizzes by.

After an hour of traveling, you seem to have returned to your start. Linda, the town supervisor, is waving to you.

"How did it go?" she says, smiling with tears in her eyes.

You look into her eyes. "I'm not quite sure." You wave to the smiling crowd and lean your bike against a wooden bench. Everyone seems normal. Reality makes sense.

But then you look down at the newspaper in Linda's hand. You can't read it. The letters seem backwards.

You reach for a pen and Linda comments that you now seem to be left-handed. But you've been right-handed for your entire life!

The town board convenes a meeting to discuss what may have happened to you, and several other brave people vow to travel down Main Street to see if they can understand the problem and the actual shape of the road and their town.

When the courageous bicyclists return a few hours later, some strange things have occurred. The bikers that started out right-handed are now left-handed, and, even weirder, physicians discover that the bikers' hearts are on the right side of their bodies instead of the traditional left! Each person comes back as a mirror image of his former self!

Gradually, more people make the trip, so that New Devonshire soon becomes a land inhabited by a peculiar mix of people and mirror people. What a nightmare this is for the local surgeons. And now couples in mixed marriages (normal people and mirror people) wonder what their children will be like. Watchmakers must design their clocks in two varieties, mirror and traditional, so that both classes of people may be comfortable and able to read the time.

This story illustrates what it might be like to live in a universe shaped like a Möbius strip. By way of analogy, if a two-dimensional "flatlander" lived in a Möbius world, he could easily "flip" himself by moving his body along his universe without ever leaving the plane of his existence. If a flatlander travels completely around the Möbius strip and returns, he will find that all his organs are reversed (figure 6.1). A second trip around the Möbius cosmos would return him to his normal orientation.

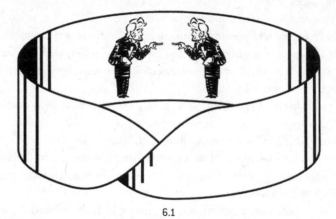

6.1
Dr. Möbius, represented as a two-dimensional human in a Möbius strip universe. If Dr. Möbius travels around the strip, his internal organs will be reversed.

When we imagine a flatlander traveling in a surface as illustrated in figure 6.1, we are considering him as a two-dimensional creature traveling *within* the surface, not on top of it. Obviously, an ant crawling on a paper Möbius strip won't be mirror reversed when traveling around it. To achieve the required effect, you might imagine an infinitely flat ant traveling in the plane of the strip.

A Möbius band is an example of a "nonorientable space." This means, in theory, it is not possible to distinguish an object on the surface from its reflected image. The surface is considered nonorientable if it has a path that reverses the orientation of creatures living in the surface. On the other hand, if a space preserves the handedness of an asymmetric structure, regardless of how the structure is moved about, the space is called "orientable."

We still need to learn more about the large-scale structure of our universe before we can determine whether orientation-reversing paths exist. Imagine the possibilities if these paths were discovered! When you traveled in a rocket ship and returned as your mirror image, all your screws, scissors, fonts, body organs, and clocks would have changed their orientations relative to your friends who never risked the journey. If your spouse or loved one returned to you reversed, would your feelings for him or her change? Would you notice the difference? Could he or she still drive your car, write in a manner legible to you, use a computer keyboard, digest the same foods, or read your books? Would the enantiomorphic molecules in the bodies of these mirror people also be reversed and unable to digest the biomolecules in our world? Would there be any advantages to being reversed? Would baseball managers with secret access to the orientation-reversing path send their players through to confuse their opponents on the playing field? Would future societies seeking uniformity send out left-handed people in space ships so that upon their return they would be right-handed? Would governments purposefully send out people to be reversed, thereby creating whole new segments of the population that could not mate with "normal" people or contract deadly pathogens that evolved to prey upon biomolecules with particular enantiomorphic characteristics?

The New Devonshire philosophers ponder all the conundrums arising in a quaint town where a bike ride converts people from one orientation to the other. As the bicyclists travel along the orientation-reversing path, *when* do their hearts change sides? What about the people the riders pass during their journeys? Which of *these* people have left-sided hearts, and which have right? Is there a sudden change in handedness, or is the change gradual? Which people along the route have the proper enzymes to permit digestion of left- and right-handed molecules?

Alan Moore in "The New Traveler's Almanac" retells part of the story of Lewis Carroll's *Through the Looking Glass*, and describes a macabre account of young Alice's eventual fate.

> [Alice] re-emerged from the strange portal flickering above the mantelpiece, which closed not long thereafter. However, in this instance there were complications. The child's hair-parting was now worn on the other side, and on examination it appeared that the positions of the organs in her body had been quite reversed. Apparently in consequence of this, Miss A. L. could no longer down or digest her normal food, and in late November of that year was weakened unto death by this disorder.

The mirroring of the body is not simply a fictional syndrome. People afflicted with Möbius mirror disease, more frequently known as dextro-cardia with situs inversus, have the position of the heart and sometimes other internal organs reversed. These people can live a normal life, with only a slightly increased risk of congenital heart defects. A total reversal of organs, called situs inversus totalis, involves complete right-to-left reversal of the thoracic and abdominal organs. People afflicted with situs inversus totalis appear in mirror image when x-rayed. Situs inversus totalis has been estimated to occur once in about seven thousand births.

Sometimes, mirrored organs can have more egregious consequences. In 2004, a baby girl was born in China's Henan province with most of her organs positioned on the "wrong" side of her body. The child's condition was not discovered until she was six months old and went for a routine checkup. Doctors found that her heart, which should lie in the left side of her chest, was on the right. Sadly, her heart was also malformed, and even the location of the atriums and ventricles in her heart were reversed. Her stomach, which should lie on the left side, was on the right, and her liver, supposed to be on the right, was on the left. Normally, a person's left lung has two parts and the right has three, but this was reversed in the Henan baby. Physicians performed heart surgery to correct the malformations, but they did not operate on her other organs, which were functioning well.

Hyperspace and Intrinsic Geometry

In the previous section, we discussed how the Möbius strip can mirror-reverse a 2-D creature embedded within its surface. But how could this possibly apply to us in our higher-dimensional universe?

Imagine alien creatures, shaped like amoebae, wandering along the surface of a large beach ball. The inhabitants are embedded in the surface, like microbes floating in the thin surface of a soap bubble. The aliens call their universe "Suibom." To them, Suibom appears to be flat and two-dimensional partly because Suibom is large compared to their bodies. However, Einsteinoid, one of their brilliant scientists, comes to believe that Suibom is really finite and curved in something he calls the third dimension. He even invents two new words, "up" and "down," to describe motion in the invisible third dimension. Despite skepticism from his friends, one day Einsteinoid kisses his wife's pseudopod and begins a long journey along what seems like a straight line around his universe. A week later, he returns to his starting point, thereby proving that his universe is curved in a higher dimension. During Einsteinoid's long trip, he doesn't feel as if he's curving, although he is curving in a third dimension perpendicular to his two spatial dimensions. Einsteinoid even discovers that there is a shorter route from one place to another. He tunnels through Suibom from point A to point B, thus creating what physicists call a "worm hole." Later, Einsteinoid discovers that Suibom is one of many curved worlds floating in 3-space. He conjectures that it may one day be possible to travel to these other worlds.

Now suppose that the surface of Suibom were crumpled like a sheet of paper. What would Einsteinoid and his fellow amoeboid aliens think about their world? Despite the crumpling, the amoebae of Suibom might conclude their world was perfectly flat because they lived their lives confined to the crumpled space. Their bodies would be crumpled without them knowing it.

This idea of curved space is not as zany as it may sound. Georg Bernhard Riemann (1826–1866), the great nineteenth-century geometer, thought constantly about these issues, and he profoundly affected the development of modern theoretical physics, providing the foundation for the concepts and methods later used in relativity theory. Riemann replaced the two-dimensional world of Suibom with our three-dimensional world crumpled in the fourth dimension. It would not be obvious to us that our universe was warped, except that we might feel its effects. Riemann believed that electricity, magnetism, and gravity are all caused by crumpling of our three-dimensional universe in an unseen fourth dimension. If our space were sufficiently curved like the surface of a sphere, we might be able to determine that parallel lines can meet (just as do longitude lines on a globe), and the sum of angles of a triangle can exceed 180 degrees (as exhibited by triangles drawn on a globe).

Around 300 BC, Euclid told us that the sum of the three angles in any triangle drawn on a piece of paper is 180 degrees. However, this is true only on a flat piece of paper. On a spherical surface, you can draw a triangle whose angles are *each* 90 degrees! To verify this, look at a globe, and trace a line along the Equator, then follow a meridian of longitude down to the South Pole, and then make a 90-degree turn and go back up another meridian of longitude to the equator. You have formed a triangle in which each angle is 90 degrees. You can also draw triangles whose angles exceed a sum of 180 degrees.

Let's return to our 2-D aliens on Suibom. If they measured the sum of the angles in a small triangle, the sum could be quite close to 180 degrees, even in a curved universe; however, for large triangles, the results could be quite different because the curvature of their world would be more apparent. The geometry discovered by Einsteinoid of Suibom would be the *intrinsic* geometry of the surface. This geometry depends only on measurements made along the surface. In the mid-nineteenth century, there was considerable interest on our own world in non-Euclidean geometries; spherical geometries in which parallel lines can intersect, for example. When physicist Hermann von Helmholtz (1821–1894) wrote about this subject, he had readers imagine the difficulty a two-dimensional creature would encounter moving along a surface as it tried to understand its world's intrinsic geometry without the benefit of a three-dimensional perspective revealing the world's curvature properties all at once. Bernhard Riemann also introduced intrinsic measurements on abstract spaces and did not require reference to a containing space of higher dimension in which material objects were "curved."

The *extrinsic* geometry of Suibom depends on the way the surface sits in a high-dimensional space. As difficult as it may seem, it is possible for Suibom creatures to understand their extrinsic geometry just by making measurements along the surface of their universe. In other words, a Suibom creature could study the curvature of its universe without ever leaving the universe—just as we can learn about the curvature of our universe, even if we are confined to it. To show that our space is curved, perhaps all we have to do is measure the sums of angles of large triangles and look for sums that are not 180 degrees. For years, legends suggested that mathematical physicist Carl Friedrich Gauss (1777–1855) attempted this experiment by shining lights along the tops of mountains to form one big triangle; however, Gauss did these kinds of experiments for triangulation and surveying purposes, and he would have known that the triangles produced by rays of light summed to 180 degrees as far as it was possible to measure. We still don't know for sure whether parallel lines intersect far away in our universe, but we do know that light rays should not be used to test ideas on the overall curvature of space because light rays are deflected as they pass nearby massive objects. This means that light bends as it passes a star, thus altering the angle sums for large triangles. However, this bending of starlight also suggests that we may imagine that pockets of our space are curved in an unseen dimension beyond our spatial comprehension. Spatial curvature is also suggested by the planet Mercury's elliptical orbit around the sun that shifts in orientation, or precesses, by a very small amount each year due to the small curvature of space around the sun. Albert Einstein argued that the force of gravity between massive objects is a consequence of the curved space near the mass, and that traveling objects merely follow straight lines in this curved space like meridians of longitude on a globe.

In the 1980s and '90s, various astrophysicists tried to experimentally determine if our entire universe is curved. For example, some have wondered if our 3-D universe might be curved back on itself in the same way a 2-D surface on a sphere is curved back on itself. We can restate this in the language of the fourth dimension. In the same way that the two-dimensional surface of the Earth is finite but unbounded (because it is bent in 3-D into a sphere), many have imagined the 3-D space of our universe as being bent (in some four-dimensional space) into a four-dimensional sphere called a hypersphere. Unfortunately, astrophysicists are unable to draw definitive conclusions because the experimental results contain uncertainties. More recent cosmological observations suggest that there is probably not much overall curvature to our visible universe; however, our visible universe is just a small portion of the entire universe, which could have all kinds of exotic topologies. The universe could be finite but with no boundary, just as a sphere's surface is finite but has no edge. In theory, this would mean that if we fly far through space, we would never encounter a wall that indicates space goes no further. There would be no sign that reads

You have reached the end of the universe.
Please turn around and go back home.

Just as on the Möbius strip, strange things would happen if we lived on the surface of a small hypersphere. By analogy, consider a flatlander living in a universe that is the surface of a small sphere. If the flatlander travels along the sphere, he returns to his starting point. If he looks ahead, he sees his own back. If you lived in a hyperspherical universe, you too could return to your starting point after traveling a long distance. If the hypersphere were small, you'd see your own back while looking forward. As alluded to in our discussion of extrinsic geometry, some cosmologists have suggested that our universe is actually the surface of a large hypersphere.

Through the last hundred years, scientists have speculated on the implications of our universe possessing other equally strange topologies like hyper-Möbius strips and hyperdoughnuts. For example, in 4-D space, various surfaces containing Möbius bands can be built that have no boundary, just like the surface of a sphere has no boundary. As we discussed, the boundary of a disk can be attached to the boundary of a Möbius band to form a "real projective plane." Two Möbius bands can be attached along their common boundary to form a nonorientable surface called a Klein bottle, named after its discoverer Felix Klein (figure 6.2). The Möbius band has boundaries—the band's edges that don't get taped together. On the other hand, a Klein bottle is a one-sided surface without edges. Unlike an ordinary bottle, the "neck" is bent around, passing through the bottle's surface and joining the main bottle from the inside.

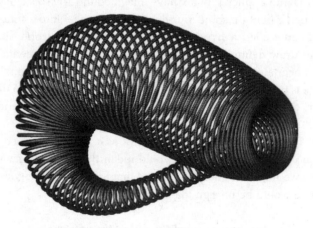

6.2

A wire-frame representation of a Klein bottle by Jos Leys. A Klein bottle is a one-sided surface. Like a Möbius strip, you can't paint the "inside" one color and the "outside" another.

We can see the interesting relationship between the Möbius strip and the Klein bottle by cutting the Klein bottle halfway along its length to form two Möbius strips (figure 6.3). One way to build an imperfect physical model of a Klein bottle in our 3-D universe is to have it meet itself in a small, circular curve. (Four dimensions are needed to create a Klein bottle without self-intersections.)

6.3
A Klein bottle sliced into two halves reveals two Möbius strips.

Imagine your frustration (or perhaps delight) if you tried to paint just the outside of a Klein bottle. You start on the bulbous "outside" and work your way down the slim neck. The real 4-D object does not self-intersect, allowing you to continue to follow the neck which is now "inside" the bottle. As the neck opens up to rejoin the bulbous surface, you find you are now painting inside the bottle.

If an asymmetric flatlander lived in a Klein bottle's surface, he could make a trip around his universe and return in a form reversed from his surroundings. Note that all one-sided surfaces are nonorientable, and if our universe were shaped like a Klein bottle, we could find paths that would cause our bodies to reverse when we returned. I urge readers to explore my book *Surfing Through Hyperspace* for additional information on higher dimensions.

I Love Klein Bottles

Let me digress from the cosmos topic and tell you about one of my favorite patents involving Klein bottles, the "One-sided beverage vessel" (U.S. Pat. 6,419,111, issued 2002), invented by Erl E. Kepner (figure 6.4). As previously noted, a Klein bottle is similar to a Möbius strip in that it has only one surface. A true Klein bottle cannot be constructed in our normal 3-D universe; however, the basic form is embodied in this invention. The Kepner Klein bottle coffee mug permits liquid from the interior of the coffee mug to exit at the bottom when suction is applied to the bottom of the cup. Kepner says,

> The beverage vessel can be used in applications where it is advantageous to be able to empty the container contents without pouring the contents over the lip of the container. An example of the possible need for this would be if an aircraft pilot's beverage vessel would need to be emptied rapidly due to air turbulence. In normal use, the pilot would handle and drink from the vessel just as with any other coffee cup.

The liquid may be drained from the cup if a vacuum is applied to the bottom of the cup, which contains an opening.

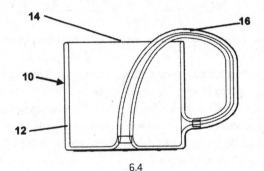

6.4

U.S. Pat. 6,419,111, "One-sided beverage vessel," by Erl E. Kepner, July 16, 2002.

Kepner concludes his patent, "The future marketing of the beverage container of this invention will use these sorts of interesting points to stimulate interest among technically well educated people and everyday people with an innate curiosity and appreciation for the wonder and beauty of mathematics and nature."

Acme Klein Bottle (www.kleinbottle.com) sells Klein bottle coffee mugs with hollow handles (figure 6.5). Astrophysicist Cliff Stoll, who heads Acme Klein Bottle, remarks, "A Klein Bottle that delivers liquid straight to your waiting lips. Yes—you heard me right. You can drink right from this cup. Pour in beer and it's a Klein Stein. Would you believe Einstein's Klein Stein? It's a true genus-1 manifold . . . with zero volume and nonorientable."

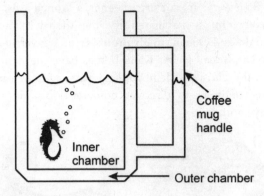

6.5 Schematic drawing of a cross section through a coffee mug Klein bottle with hollow handle, sold by Acme Klein Bottle (www.kleinbottle.com).

The Acme Klein Bottle Web site continues to extol the cup's virtues. With Acme's Klein bottle mug, you can fill the inside with coffee and the outside with tea. The handle connects the inner and outer chambers, providing a topological hole. The outer chamber (which is topologically the inner chamber) insulates the inner chamber (which topologically is also the outer chamber). The seven millimeter air space separates the inside from the outside and keeps cold drinks cold longer and hot drinks warm longer. Stoll writes, "This Klein Stein is ideal for the mathematical physicist who needs a glass of water while accepting her Nobel Prize."

Together with Toronto's Kingbridge Centre and Killdee Scientific Glass, Cliff Stoll has created the world's largest glass Klein bottle. The Kingbridge Klein bottle is 1.1 meters tall, 50 centimeters in diameter, and is made of 15 kilograms of clear Pyrex glass (or 42 inches tall, 20 inches across, and 35 pounds.) Cliff Stoll remarks,

One sided, boundless, and mathematically nonorientable. It tickles topologists and amazes visitors. It's the size of a five-year-old child. A ferret can climb "into" it. It's been a nontrivial glass-blowing project. Indeed, very few glassblowing shops could handle this job. (One glass blower said, "Too scary for us!")

Not content with your everyday Klein bottles, mathematicians and computer artists enjoy exploring related shapes with odd properties. Figures 6.6 and 6.7 depict Bonan-Jeener's Klein surface and a Jeener's Klein surface of the second order, as rendered by computer artist Jos Leys. The surfaces get their name from Patrice Jeener, a French artist and copper-plate engraver who is fascinated by the theory of surfaces, and Frenchman Edmond Bonan, math professor at Université de Picardie Jules Verne. The Bonan-Jeener Klein bottles have the same topological properties as the classical Klein bottle. Jeener, although self-taught in mathematics, continues to discover equations for odd surfaces that delight the eye and mind.

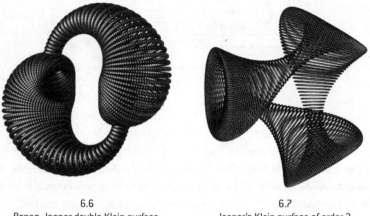

6.6
Bonan-Jeener double Klein surface.
[Rendering by Jos Leys.]

6.7
Jeener's Klein surface of order 2.
[Rendering by Jos Leys.]

The Banchoff Klein bottle (figures 6.8 and 6.9) is also based on the Möbius band. The computer algorithm I used to produce this form is outlined in the reference section for this chapter. Powerful computer graphics applications allow us to design unusual objects such as these and then investigate them by projecting them in a 2-D image.

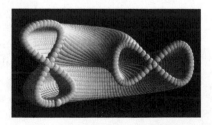

6.8
Banchoff Klein bottle. [Rendering by the author.]

6.9
Cross section of Banchoff Klein bottle, revealing "internal" surfaces.
[Rendering by the author.]

If you are a teacher, have your students design and program their own patterns by modifying the parameters in the equations in the reference section, and make a large mural of all the student designs labeled with the relevant generating formulas. In the last decade, even serious mathematicians have begun to enjoy and present bizarre mathematical patterns in new ways—ways sometimes dictated as much by a sense of aesthetics as by the needs of logic. Moreover, computer graphics allow nonmathematicians to better appreciate the complicated and interesting graphical behavior of simple formulas.

To produce the object in figure 6.8, I place spheres at locations determined by formulas that are implemented as computer algorithms. Many of you may find difficulty in drawing shaded spheres; however, quite attractive and informative figures can be drawn simply by placing dots at the x, y, z locations.

Hyperspace Mirrors

In chapter 5 we discussed how a Möbius strip comes in two forms: the right-handed and left-handed. One form could only be turned into the other if we could rotate it in the fourth dimension. Although the vague notion of a fourth dimension had occurred to mathematicians since the time of Kant, most mathematicians dropped the idea as fanciful speculation with no possible value. They had not discussed the fact that an asymmetric solid object could, in theory, be reversed by rotating it through a higher space. It was not until 1827 that Möbius showed how this could be done—eighty years after Kant's papers on dimension.

What does it mean to rotate an object in a higher dimension? If you encountered a flatlander, you could, in principle, lift him out of his plane and flip him around. As a result, his internal organs would be reversed. For example, a heart on the left side would now be on the right. Similarly, a 4-D being might flip us around and reverse our organs. Although such powers are possible within the auspices of hyperspace physics, I should remind readers that the technology to manipulate space in this fashion is not possible; perhaps in a few centuries we will explore hyperspace in ways only dreamed about today in science fiction.

Many creatures in our world, including ourselves, are bilaterally symmetric in their exterior form; that is, their left and right sides are similar, like the harlequin longhorn beetle in figure 6.10. On each side of our bilaterally symmetric body is an eye, ear, nostril, nipple, leg, and arm. In 2004, paleontologists at the Nanjing Institute of Geology and Palaeontology discovered the oldest example of bilateral symmetry in a rock quarry in southern China. The *Vernanimalcula guizhouena*, a microscopic creature that lived on the sea floor six hundred million years ago, has paired digestive canals on either side of the gut.

6.10
The harlequin longhorn beetle is a classic example of bilateral symmetry.

One way to visualize the flipping of objects in a higher space is to consider the squiggly blobs in figure 6.11, which are obviously not bilaterally symmetric. They make an enantiomorphic pair because they are congruent but not superimposable without lifting one out of the plane. We've discussed enantiomorphs when discussing molecular Möbius strips. Similarly, in our three-dimensional world, there are many examples of enantiomorphic pairs—these consist of asymmetric solid figures such as your right and left hands. (If you place them together, palm to palm, you will see each is a mirror reflection of the other.) The squiggly blobs in figure 6.11, like your two hands, cannot be superimposed, no matter how you rotate and slide them on the plane. However by rotating the blobs around a line in space, we can superimpose one blob on its reflected image. Similarly, your own body could be changed into its mirror image by rotating it around a plane in 4-space.

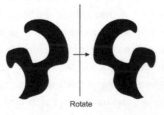

Rotate

6.11
The squiggly shape at the left can be superimposed on the squiggle at the right only if it is first rotated out of the page into a higher dimension.

Mirrors in our 3-D world are two-dimensional planes. In four dimensions, figures are mirrored by a solid. Mirrors are always one dimension less than the space in which they operate.

If there were a hyperperson in 4-space looking at our right and left hands, to him they would be superimposable because he could conceive of rotating them in the fourth dimension. The same would apply to Möbius strips with different handedness as well as seashells with clockwise and counterclockwise spirals.

Oddly enough, there are many examples of non-enantiomorphic hands on people and gods in ancient artwork. I have never been able to ascertain why this is so. Surely, the artists had the capability and insight to see that our left and right hands are mirror images. If you look closely, Egyptian wall carvings and paintings frequently depict pharaohs with two left hands. This collapse of enantiomorphism also occurs in Mesopotamian art, in depictions of the Babylonian god Marduk, for example.

Möbius Worlds

If our entire universe were suddenly changed into its mirror image, would we perceive a difference? To answer this question, consider "Lineland" inhabited by only three intelligent gazelles: "Gazelle 1," "Gazelle 2," and "Gazelle 3," all facing east; that is to say, they are all looking to the right (figure 6.12). Although the diagram shows them as two-dimensional drawings, assume that they are really one-dimensional and cannot leave the line, which is their universe and in which they are embedded. If we reverse Gazelle 2, the change will be apparent to Gazelle 1 and Gazelle 2. But if we reverse the entire line of Lineland, the one-dimensional gazelles would not perceive a change. We higher-dimensional beings would notice that Lineland had reversed, but that is because we can see Lineland in relation to a world outside it. Only when a *portion* of their world has reversed can they become aware of a change. The same would be true of our world. In a way, it would be meaningless to say our entire universe was reversed because there would be no way we could detect such a change. Why is our world a particular way?

6.12
If our entire universe were suddenly changed into its mirror image,
would we perceive a difference?

Philosopher and mathematician Gottfried Wilhelm Leibniz (1646–1716) believed that to ask why God made the universe this way and not another is to ask "a quite inadmissible question." To get a better understanding of Leibniz's comment, consider a two-dimensional Flatland teeming with intelligent amoebae. To mirror-reverse the entire Flatland universe, all we have to do is turn the plane over and view it from the other side! In fact, we don't even have to turn the world over. Consider Flatland to be like a vertical ant farm in which the ants are essentially confined to a 2-D world. The world is a left-handed world when viewed from one side of the glass and a right-handed world when viewed from the other. In other words, Flatland does not have to change in any way when you view it from one side or the other. The only change is in the spatial relation between Flatland and an observer in 3-space. In the same way, a hyperbeing could change his position from a 4-D "up" to a 4-D "down" and see a seashell with a right-handed spiral become a left-handed spiral. If he could pick up the shell and turn it over, it would be a miracle to us. What we would see is the shell disappear and then reappear as its mirror image. All of this means that enantiomorphic structures are seen as identical and superimposable by beings in the next higher dimension. Perhaps only a god existing in infinite dimensions would be able to see all pairs of enantiomorphic objects as identical and superimposable in all spaces. Many other kinds of spatial distortions are discussed in detail in my book *Surfing Through Hyperspace*.

The Three-Torus and Other Magnificent Manifolds

Both the Möbius band and the Klein bottle are surfaces or, as mathematicians call them, manifolds. More precisely, any object that is nearly "flat" on small scales is a manifold. For example, a sphere is nearly flat if we magnify a tiny portion of the surface, and this is why some people centuries ago believed the Earth was flat. Close up, the Earth does indeed look flat, although the ancient Greeks noticed that a ship's mast was the last part of a ship to disappear over the horizon.

The surface of a sphere is two-dimensional, but manifolds can have any dimension. A smooth line, even if it curves, is a 1-D manifold because a tiny section of the curve looks like a line. The curve has the topology of a line. Similarly, a two-manifold has the local topology of a plane. A three-manifold has the local topology of three-dimensional space.

As we have been discussing, a nonorientable manifold has a path that brings a traveler back to his starting point mirror-reversed. The surface of a sphere is an orientable manifold. No path exists on a sphere that will mirror-reverse a creature traveling in its surface.

On a sphere, it's also possible to travel in any direction and return to your starting point. You can't fall off the edge of a sphere's surface. Thus, a sphere is an example of a manifold with no boundary. The surface of a torus or doughnut is another orientable surface with no boundary. On the other hand, imagine a cylinder made by taping the right and left side of a piece of paper together. This cylinder has two boundaries, one at the top edge and one at the bottom.

Please allow me a little repetition to help you visualize some of these new concepts. Recall that the Möbius strip has a boundary just like the cylinder, but it only has one boundary, not two. You can verify this for yourself by trying to color one edge of a Möbius paper strip red and the other blue. You can't. If you trace along the edge you'll eventually return to your starting point. As discussed, using the Möbius strip as a starting point, we can create another nonorientable surface, eliminating the edge by curving the strip so that the "apparent" two edges are joined. Alas, you can't really construct the new manifold in our 3-D world because it would intersect itself. A higher-dimensional creature could, however, make the closed surface, which is called a Klein bottle. All we can do in our 3-D world is make a model for a Klein bottle which shows a self-intersection at the bottle's neck. This is a projection of a 4-D object into 3-space, just like a circle is a projection of a sphere. Unlike the Möbius strip, an actual Klein bottle would have no boundary.

As mentioned in chapter 5, another fascinating nonorientable surface is the real projective plane. We can imagine constructing it by looking at the square starting shape in figure 5.14 for the Klein bottle and twist the top and bottom edges in addition to the left and right edges before gluing. Another way to model the projective plane is to imagine a hemisphere and connect each point on the rim to its corresponding point on the opposite side, but with a twist (figure 6.13).

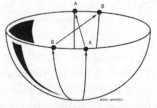

6.13

The projective plane may be visualized by imagining a hemisphere and connecting each point on the rim to its corresponding point on the opposite side, but with a twist.

The projective plane is a one-sided surface, like a Möbius strip, but it cannot be realized in three-dimensional space without crossing itself. Like the Klein bottle, the projective plane has no boundary.

In general, a finite manifold without boundaries cannot be built in the same dimension in which its inhabitants live, but it's easy to extend our imagination to higher dimensions by reasoning, using analogies of lower dimensions. You can imagine taking a piece of paper (which is a two-manifold with a boundary on all sides) and making a torus simply by connecting the right and left sides of the paper and the top and bottom. To do this, you'd have to fold the 2-D paper in three space. Similarly, we can try to imagine making the same kind of connection with a solid cube, which is a three-manifold with a boundary on all sides. Imagine if we could stretch the cube to connect the right wall to the left wall. If you existed in such an object, you could toss a ball to the right and it would roll out at the left if the distance it would have to travel was not too far. Now try to imagine connecting the cube's front wall to its back wall and its top wall to its bottom wall. If you could perform this kind of connecting in a higher dimension, you could create a new manifold called a *three-torus*. This object has no boundary, like a two-torus, and if you lived in a three-torus universe, it would seem to be an infinite space.

Over the years, scientists have suggested all kinds of hypothetical shapes for the universe in which we live, including a three-torus. If we lived in a three-torus, you could peer out into the universe with a powerful telescope and could, in theory, be gazing in the direction of your own back.

We could also visualize constructing a nonorientable three-manifold by recalling that a Möbius strip can be made by connecting the ends of a rectangle after we give it a half twist or flip. If we could connect the front and back side of a cube after giving one face a half twist, when you walked toward the back, you would eventually come out the front mirror-reversed. All kinds of crazy universes could be made by connecting various walls of the cube with different partner walls, with or without twists.

Scientists continue to ponder the shape of the universe. According to Charles Seife in "Polyhedral Model Gives the Universe an Unexpected Twist," a team of scientists from France and the United States have studied measurements from the Wilkinson Microwave Anisotropy Probe (WMAP) satellite and reached a surprising conclusion: the universe might be finite and twelve-sided. Although most astronomers with whom I have talked look on this idea as an exotic possibility rather than a mainstream theory, I don't think it has been ruled out by the data. According to this model, opposite faces of the dodecahedron correspond in unusual ways to each other. In fact, these faces are actually the same face so that a spaceship flying out one side of the universe winds up flying back into the other side.

To make a finite dodecahedral space, one would glue together opposite faces of a slightly curved dodecahedron—a shape like a soccer ball with twelve pentagonal sides. Of course, such gluing is difficult to imagine in our ordinary 3-D space.

In the spirit of full disclosure, I should note that scientists' theories about the shape of the universe change almost every month. In April of 2004, Frank Steiner at the University of Ulm in Germany suggested that the universe is shaped like a medieval horn—a very long funnel. In Steiner's model of the universe, technically known as a Picard topology, the universe is infinitely long in the direction of the funnel's spout, but so narrow the universe has finite volume.

Multiple Universes

Today, many of my physicist colleagues ponder big questions, like the formation of the universe and the ultimate shape of space. Many cosmologists have suggested that the big bang that created our universe is just one of many big bangs. Luckily for us, our big bang produced stars and planets. Most of the planets in our universe are dead worlds, but Earth is notable because it has conditions on which life can evolve. Similarly, most of the other universes produced by the big bangs might be dead universes because they did not happen to have the conditions that permitted stars to shine. Just in the last few decades, increasing numbers of cosmologists are starting to accept this idea of multiple universes, that is, the multiverse, in part due to superstring theory, which suggests that many forms for a universe are possible.

If an infinite number of random (non-designed) universes exist, ours could be just one that permits carbon-based life. Some researchers have even speculated that child universes are constantly budding off from parent universes and that the child universe inherits a set of physical laws similar to the parent, a process reminiscent of the evolution of biological characteristics of life on Earth.

The universes that are "successful" from a cosmological-Darwinian perspective are those that produce large numbers of child universes with long lifetimes. For example, if we suppose that the central singularities in black holes produce other universes, as some have suggested, universes with numerous black holes will be successful. Because many forms of black holes take a long time to form, these universes will be sufficiently long-lived to allow for galactic formation and stellar nucleosynthesis (the formation of elements in stars that life needs). This means that successful universes automatically have nearly the right characteristics for the appearance of life forms (figure 6.14). To put it another way, as the cosmological ecosystem evolves, the most common universes are those which produce large numbers of black holes, stars, and life-forms. If the speculative scenario of evolving universes describes reality, then our universe may not be unusual.

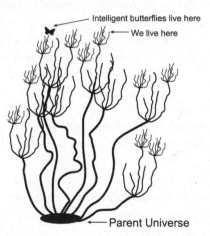

6.14 Cosmological Darwinism. Researchers postulate that baby universes are spawned from parent universes, and the babies also have babies. The children inherit similar physical laws from their parents, and "successful" universes have a tendency to produce successful offspring. Successful universes are long-lived and have many children and stars, all of which encourages the formation of biological life.

If our universe is infinite, as some cosmologists suggest, our visible universe is just a tiny portion of the cosmos. What we call the laws of nature may be just the laws in our pocket of the cosmos, and other laws may reign elsewhere. If our universe is infinite, it is likely that configurations of atoms, by chance alone, mimic those in our visible universe. According to astrophysicist Max Tegmark, the number of meters one must travel to find an exact copy of yourself, assuming that the universe is homogeneous and infinite, is $10^{10^{29}}$. In other words, by the laws of chance alone, a configuration of atoms just like yours exists somewhere else in an infinite universe.

Here's another way to look at this. We live in a visible universe easily encompassed by a sphere one hundred billion light-years across, with a finite number of configurations for the matter and energy contained within. Let's imagine our visible universe as a gigantic bubble floating within our larger universe. (We cannot see infinitely far because the universe has a finite age and because information cannot travel faster than the speed of light.) If our universe is infinite, as some modern physicists believe, then nearly identical copies of our bubble likely exist, and they contain a replica of our Earth and of you. According to physicist Max Tegmark, on average, the nearest of these identical bubbles is about 10 to the 10^{100} meters away. Not only are there infinite copies of you, there are infinite copies of variants of you. It is almost certain that right now you have red eyes and are kissing someone who speaks Etruscan with long fangs in some other bubble. If we accept the notion of an infinite universe—which is suggested by modern theories of cosmic inflation—infinite copies of you exist, altered in fantastically beautiful and ugly ways. If you yearn for some lover you can never have in this world, it is almost certain you are with this person somewhere else in this universe. Be happy.

We Are Simulations

In our own region of the universe, we've already developed computers and the ability to simulate lifelike forms using these computers and mathematical rules. I believe that one day we will create thinking beings that live in rich simulated ecosystems. We'll be able to simulate reality itself, and perhaps more advanced beings are already doing this elsewhere in the universe. Huge supercomputers would have the capacity to simulate not just a tiny fragment of reality, but a substantial fraction of an entire universe.

What if the number of these simulations is larger than the number of universes? Could we be living in such a simulation? Astronomer and philosopher Martin Rees suggests that if the simulations outnumber the universes, "as they would if one universe contained many computers making many simulations," then it is likely that *we* are artificial life. He notes that this theory allows for "virtual time travel" because the advanced beings who create the simulation can rerun the past. Rees says the following in his essay "In the Matrix" (or "Living in a Multiverse"):

Once you accept the idea of the multiverse, and that some universes will have immense potentiality for complexity, it's a logical consequence that in some of those universes there will be the potential to simulate parts of themselves, and you may get a sort of infinite regress, so we don't know where reality stops and where the minds and ideas take over, and we don't know what our place is in this grand ensemble of universes and simulated universes.

Astronomer Paul Davies in "A Brief History of the Multiverse" has made similar observations.

Eventually, entire virtual worlds will be created inside computers, their conscious inhabitants unaware that they are the simulated products of somebody else's technology. For every original world, there will be a stupendous number of available virtual worlds—some of which would even include machines simulating virtual worlds of their own, and so on ad infinitum.

Some readers may not be aware of the strides computer scientists and biologists have already made in the area of "artificial life," in which life-like entities with complex behaviors are simulated using simple rules implemented in computer software.

The study of artificial life reminds me of mycology (the study of fungi) or myrmecology (the study of ants). Researchers have simulated simple life-forms with short life spans, living in simple societies. Professor Tom Ray of the University of Oklahoma's department of zoology created Tierra, a system in which self-replicating machine code programs evolved by natural selection. Although these creatures were very small, only a few instructions long, they exhibited many behavioral patterns found in nature. Diverse ecological communities emerge when many of these and other simulated biomorphic entities interact. These kinds of digital communities have been used to experimentally examine ecological and evolutionary processes, including host-parasite population regulation, the effect of parasites in enhancing community diversity, evolutionary competition, punctuated equilibrium, and the role of chance in evolution.

Other kinds of natural behaviors have been exhibited by Craig Reynolds's "Boids"—entities that flock or school like birds and fish. Craig, who now works for Sony Computer Entertainment, used only three simple rules to govern the life of the Boids: 1) steer to avoid getting too close to neighbors; 2) steer to keep on the average heading of the flock; and 3) steer to stay near the average position of the neighbors. Rules about goal seeking and obstacle avoidance can be added to allow the artificial creatures to navigate through a world filled with objects. The resulting behaviors based on these simple rules are remarkable, and it is not hard to imagine that simple rules and simulations known as cellular automata could develop complex societies of reproducing entities, especially when millions or billions of the creatures interact in huge worlds.

Examples of other artificial life-forms include Kevin Coble's Neoterics, Larry Yeager's Polyworld, and Karl Sims's multilimbed creatures (MCs) that compete with one another. The MC brains are neural nets and have several sensors. Through competition, the creatures evolve intelligent behaviors that would be hard for humans to actually design and build into them.

The Avida system—a joint project of the Digital Life Laboratory at the California Institute of Technology and the Microbial Evolution Laboratory at Michigan State University—provides a software platform in which digital organisms breed thousands of times faster than common bacteria and shed light on some of the biggest unanswered questions of evolution. Darwin-at-Home is a planetwide effort to create networked digital ecosystems and to recreate the evolution of life on Earth by permitting computer creatures to evolve (www.darwinathome.org). The Darwin teams hope to observe lifelike evolutionary processes in a virtual or robotic space. Their interactive computational platform is distributed across a large pool of networked computers, which allows people to shape each digital biotic ecosystem.

One of the most famous and earliest cellular automata life-forms is John Horton Conway's game called Life. In this simple simulation, cells live or die on a two-dimensional grid of cells when they follow just two rules: 1) A cell is turned on (lives) if three of its neighbors are turned on, and 2) A cell remains on if two or three of its neighbors are also on; otherwise it is turned off (dies). These simple rules control the birth, survival, and death of any cells through time. Sometimes, entities or shapes composed of a collection of cells evolve and move around the checkerboard universe while maintaining their overall shape, just like a creature moving though a pond. In fact, some forms evolve that are able to maintain their shape and spawn other shapes that are then able to "explore" the environment, essentially simulating the act of reproduction. If lifelike phenomena emerge using such simple rules, we can expect that certain rules executed on checkerboard universes could spawn complex societies given sufficient time and a sufficiently large world on which to evolve.

Science fiction author Greg Egan suggests in *Permutation City* that medical imaging technology will improve, and that by 2020 it will reach the point where individual neurons can be mapped and the properties of individual synapses measured noninvasively. "With a combination of scanners, every physiologically relevant detail of the brain could be read from the living organ—and duplicated on a sufficiently powerful computer."

Egan suggests that at first only isolated neural pathways will be modeled: "portions of the visual cortex of interest to designers of machine vision, or sections of the limbic system whose role had been in dispute."

These fragmentary neural models yielded valuable results, but a functionally complete representation of the whole organ . . . would have allowed the most delicate feats of neurosurgery and psychopharmacology to be tested in advance. . . . In 2024 . . . a Boston neurosurgeon ran a fully conscious Copy of himself. . . . The first Copy's first words were: "This is like being buried alive. I've changed my mind. Get me out of here."

Measuring the Universe's Shape

In a previous section, we discussed a NASA satellite known as the Wilkinson Microwave Anisotropy Probe that has provided information that allows scientists to further speculate on the shape of the universe. The satellite records the universe's pattern of heat in the form of faint microwave radiation. This radiation is an "afterglow" of the big bang itself, and thus paints a portrait of the early universe. If the universe were infinite, the remnants of the big bang should appear randomly around the sky at all sizes. But, according to the satellite's data, the wave size may be limited, with no waves extending more than sixty degrees across the sky. If the universe were a symphony, it would be missing its deepest notes, produced by the cello, bass, tuba, and bassoon. What are we to make of these missing notes? Perhaps they indicate that the universe is finite and thus cannot produce waves larger than itself. In such a universe, astronauts could conceivably travel into space in one direction and end up finally returning to their starting points like caterpillars crawling on the surface of a ball.

Dr. George Efstathiou of Cambridge University believes that the Wilkinson satellite data may be consistent with a hypersphere. In this instance, fluctuations larger than the radius of the sphere might be dampened, producing the observed cutoff in the radiation pattern. Also, the universe could be spherical yet so large that the visible universe we observe seems flat just like the Earth around us may seem flat, because it's just a small patch on a giant sphere.

If the universe were finite in at least one direction, like a cylinder or doughnut, the background radiation pattern would have certain kinds of restrictions in those directions. Some researchers have proposed that the universe could have been born as a doughnut shape. In a doughnut universe, which is also an example of a multiply connected universe, light can travel from point A to point B by more than one direct path. Scientists I interviewed are divided as to whether the universe is finite or infinite. Some say that an infinite universe is most likely, and, as discussed, in these universes, almost anything can happen, including there being multiple replicas of each one of us, but with slight changes, like some replicas having horns on their heads. Other scientists say that nature has an easier time making a finite universe. (Examples of finite spaces include surfaces known as "compact manifolds," such as the circle, the n-dimensional sphere, and the torus. The term "compact manifold" usually implies a shape that is closed and doesn't have a boundary.) According to Dennis Overbye in "Universe as Doughnut: New Data, New Debate," a very likely, and perhaps the simplest, shape for a compact, finite universe is a 3-torus, a doughnut wrapped in three dimensions. We have already described this as a cube whose opposite sides are somehow glued together. Or think of a computer screen in which you move your cursor off the top of the screen only to have it wrap to the bottom of the screen, and cursors moving to the right reappear at the left just as they leave the screen. A 3-torus universe is considered a *flat* universe in the mathematical sense because, for one reason, the angles of a triangle on its surface sum to the usual 180 degrees, as if drawn on a plane of paper. This is not true for a triangle on a sphere. Parallel lines never meet on a plane or on a torus, but they can meet on a sphere.

Another way to think about a torus being flat is to realize that we can straightforwardly map a plane onto the torus surface. As has been well-known for centuries, a planar map cannot be mapped to a sphere without distortions, and this is why so many world map projects have been developed in the attempt to make useful maps for our spherelike world.

Perhaps the most famous of these projections in the Mercator projection, developed in 1568 by Gerardus Mercator, a Flemish geographer, mathematician, and cartographer. Alas, this projection causes considerable size distortion at latitudes approaching the poles, making Greenland look bigger than South America.

Wolves on Cylinders

If the notion of flat space and closed space is still confusing you, consider that a flat piece of paper is a model for flat closed space. Roll this paper into a cylinder. To see why this space is still flat, imagine a wolf walking within the surface of the cylinder. The space is said to be flat because the wolf does not rotate as it walks around a closed path, trying always to keep its body parallel to its previous position. The space is considered closed because a wolf walking along a cylinder eventually can come back to itself by walking in a straight line—indicating the closure of its space (figure 6.15). Similarly, a torus is a flat space. One way to visualize a torus is to start with a square, called a fundamental domain for the torus. Visualize the square as a flat piece of paper that can be rolled up by gluing its right and left sides together. A torus can be created by connecting the top and bottom of the paper cylinder together.

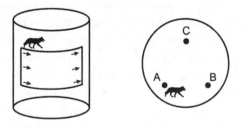

6.15

Wolves in cylindrical (left) and spherical (right) universes. The head of the wolf is represented by an arrowhead. If the wolf walks along a closed path on the cylinder, its body does not rotate. Despite superficial appearances, this space is *not* considered curved. On the other hand, the wolf on a spherical surface experiences a rotation of its body as described in the text.

Topologists call planes, cylinders, and tori Euclidean spaces. In Euclidean geometry, for each straight line and a point off the line, there is a unique line parallel to the first that passes through the point. And, as we've said, the sum of the angles of a triangle equals 180 degrees. The torus is a Euclidean (flat) 2-manifold. If the 3-D space in which we live was toruslike, we would be living in a 3-torus. However, even if we were certain that our universe was Euclidean, it could have many different forms in addition to the 3-torus. In fact, there are only eighteen possible Euclidean 3-manifolds. Of these eighteen manifolds, eight are nonorientable; that is, they contain an orientation-reversing loop. If you travel through one of these orientation-reversing paths, you might not even notice the change until you returned to Earth where all the clocks would be running counterclockwise.

Cosmologists, driven by an insatiable curiosity, scan the sky looking for hints that our universe could be nonorientable. In principle, we would observe characteristic energy patterns if we lived in such a space. So far, these patterns have not been observed.

Returning our attention to figure 6.15, let us try to better understand why the sphere's surface is a model of closed space, but it is not flat. Here's yet another way to distinguish a flat space, in which the wolf walks on a cylinder, from a curved space, in which the wolf walks on a spherical surface. On the right side of figure 6.15, the wolf attempts to walk from A to B to C to A, all the while keeping itself parallel to its previous orientation, with its head pointing towards the right of the figure. However, when it gets back to A, its head will point more in the direction toward C than to B, as it originally did. For example, as the wolf travels from B to C, its body gradually begins to point upward. (If you find this movement difficult to visualize, many interactive demos on the Web exist, such as the one at John Sullivan's page, http://torus.math.uiuc.edu/jms/java/dragsphere/.) The space is said to be curved because as the wolf walks around a closed path while always trying to keep its body parallel to its previous position, its body experiences a rotation. It has a nonzero rotation of its head. The curvature of the space is revealed by a process called *parallel transport.* On the other hand, the wolf can walk around a cylindrical surface along the path shown and return to its starting point with zero rotation of its body. So even though a cylinder's surface looks curved, it is not curved when considered as a model of space.

Spherical, Flat, and Hyperbolic Universes

Cosmologists continue to ponder various possible "shapes" for our universe. For example, space may have positive curvature and resemble the surface of a sphere. The geometry of the universe may be flat or Euclidean—or it may be hyperbolic, with a negative curvature that may be crudely visualized by examining the seat of a saddle. Many astronomers are considering a flat cosmos because it is closely tied to "inflation theory"—a popular conjecture that the universe underwent an early period of rapid expansion that amplified random subatomic fluctuations to form the current structures in our universe. In much the same way that expansion makes a small region of a balloon look flat, inflation would stretch the universe, smoothing out any curvature it might have had initially. It is astonishing that we live in an age that all of these conjectures will soon be testable with satellites scanning the universe's microwave background radiation. For example, in a hyperbolic universe, strong temperature variations in the microwave background should occur across smaller patches of the heavens than in a flat universe. (See Ron Cowen's and Ivars Peterson's 1998 *Science News* articles in the references section.) In a closed, hyperbolic universe, what astronomers might think is a distant galaxy could actually be our own Milky Way—seen at a much younger age because the light has taken billions of years to travel around the universe. Montana State University's Neil Cornish and other astronomers suggest that, "If we are fortunate enough to live in a compact hyperbolic universe, we can look out and see our own beginnings."

It is possible that the universe has a strange topology so that different parts of it interconnect like pretzel strands. If this is the case, the universe merely gives the illusion of immensity, and the multiple pathways allow matter from different parts of the cosmos to mix. In the pretzel-shaped universe, light from a given object has several different ways to reach us, so we should see several copies of the object. If our universe was a 3-torus, we'd be able to look out into space and see stars over and over again due to the wrap-around nature of the universe.

Inflation theory, which suggests that the universe underwent a runaway accelerating expansion early in time, implies that our observable universe today is a bubble 156 billion light-years in diameter. (During the inflation, what is now the observable universe blew up from a region smaller than a proton to larger than a grapefruit in a minuscule 10^{-35} of a second.) The observable universe today seems awfully big, but it may only be like a grain of dust floating in a universe trillions of light-years across. The notion of a "small" finite universe runs counter to inflation theory. And if we accept inflation theory, then we are also likely to accept the notion of multiple universes because once inflation starts, it reoccurs, spawning a chain of universes, like bubbles within bubbles. Inflation produces universes that are flat.

I'm often asked, "How could the universe be infinite if it was all concentrated into a point at the big bang?" One answer is that the universe need not have been concentrated into a point at the time of the big bang; perhaps only the observable universe was concentrated into a point. The universe could have been born infinite at the big bang. (Here, when I say "observable universe," I refer to what in principle could have been observed given the finite speed of light.)

Worlds Close to Ours

Today, many physicists suggest that there are universes parallel to ours, like layers in an onion, and that we might detect them as gravity leaks from one layer to an adjacent layer. For example, light from distant stars might be distorted by the gravity of invisible objects residing in parallel universes only millimeters away. Since 1997, scientists at the University of Colorado at Boulder have conducted experiments to search for these possible nearby universes. These researchers search for tiny deviations in Newton's inverse square law of gravity that might be caused by matter in parallel universes or in a hidden dimension. The whole idea of multiple universes is not as far-fetched as it may sound. According to a recent poll of seventy-two leading physicists conducted by the American researcher David Raub, 58 percent of physicists (including Stephen Hawking) believe in some form of multiple universe theory.

One of the latest and most mind-boggling theories of cosmogenesis suggests that all the matter and energy in our universe was created when a four-dimensional fragment of another universe wrinkled, floated through 5-D space, and then imprinted itself on our universe. Charles Seife eloquently describes what is called the "ekpyrotic model," in *Science* magazine:

> In [effectively] five-dimensional space float two perfectly flat four-dimensional membranes, like sheets drying on parallel clotheslines. One of the sheets is our universe; the other a "hidden" parallel universe. Provoked by random fluctuations, our unseen companion spontaneously sheds a membrane that slowly floats towards our universe . . . The floater speeds up and splats into our universe, whereupon some of the energy of the collision becomes the energy and matter that make up our cosmos.

Some have likened the ekpyrotic "membrane creator" from the hidden universe to the "Spirit of God" in the second line of Genesis, which reads, "Now the earth was formless and empty, darkness was over the surface of the deep, and the Spirit of God was hovering over the waters." The formless, dark, and empty earth corresponds to our universe prior to the four-dimensional membrane splat that created matter and energy here. The "hovering" corresponds to the floating of the membrane. Moreover, the ekpyrotic model suggests that at any moment another membrane could peel off, float towards our universe, and destroy us all. Some physicists say they already see signs of our impending doom presaged by the accelerated expansion of our universe. The possible doom at our doorsteps, predicted by the ekpyrotic model of our universe, also has correspondence with various biblical prophecies of apocalypse and the End of Days. Obviously, the idea of fitting theoretical physics to biblical passages involves extreme flights of fancy, but I enjoy the endless debates and mind-stretching dialogue that result.

We can go even further and think about the wild implications for multiple universes and what they say about our power in relation to God's. Stanford University physics professor Andrei Linde has speculated that it might be possible to create a new baby universe in a laboratory by violently compressing matter at high temperatures—in fact, one milligram of matter may initiate an eternal self-reproducing universe (see Rucker reference in references section). What would be the economic or spiritual gain we would get from creating a universe, considering it would be extremely difficult, if not impossible, to enter the new universe from ours? Should *we* be looking for such evidence in the values of the Planck length, pi, or the golden ratio? Would God care if we created such universes at will? Andrei Linde and writer Rudy Rucker have discussed methods for encoding a message for the new universe's potential inhabitants by manipulating parameters of physics, such as the masses and charges of particles, although this would be a precarious experiment given the difficulty of manipulating these constants in a way that both codes a message and permits life to evolve.

● Pretzel Transformation Puzzle

One of my favorite transformations, known to aficionados as the pretzel transformation, involves the conversion of the double-looped structure at the top left of figure 6.16 to the structure at the top right without cutting the loops. In other words, it is possible to unchain the two loops without breaking one of the loops. All you have to do is assume that the object is made from a very elastic material so that it can be stretched. Here's a hint: start by enlarging one of the two loops. (Turn to the solution sections for an answer.)

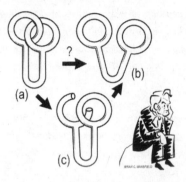

6.16

Pretzel transformation. Can you transform the linked object at the left into the unlinked object at the right without cutting one of the loops? (Obviously, you could cut the object as shown at bottom, but that would be too easy!)

Möbius Cosmos

✒ *If we lived on a hyper-Möbius strip, and we peered in front of us, we would see the back of someone's head. At first, we wouldn't think it could be our head, because the part of the hair would be on the wrong side. If we reached out and placed our right hand on his shoulder, then he would lift up his left hand and place it on the shoulder of the person ahead of him. In fact, we would see an infinite chain of people, with hands on each other's shoulders, except the hands would alternate from the left to right shoulders.*

—Michio Kaku, Hyperspace

🎞 *It may seem that the Möbius strip or the bridges of Königsberg are a world away from the cosmic connection, but that is not so. Outer space is the testing ground for Einstein's theory of curved space and time. It is among the unimaginably powerful gravitational fields of distant astronomical bodies that space is buckled and bent, perhaps even torn and joined in Möbius forms, or more complicated topologies.*

—Paul Davies, The Edge of Infinity: Beyond the Black Hole

🎞 *Now a team led by Ruth Durrer of the University of Geneva in Switzerland has an explanation [as to why we move about in a 3-D universe]. The idea is that the cosmos once included branes with up to eight dimensions, floating about at random in a nine-dimensional space. In their model, this 9-D space has the form of a torus, or doughnut, with each dimension circling back on itself*

—Stephen Battersby, "How 3-D Space Survive the Great Destruction,"
New Scientist

🎞 *Homer, your theory of a donut-shaped universe is intriguing.*

—Stephen Hawking to Homer Simpson in The Simpsons
episode "They Saved Lisa's Brain," 1999

CHAPTER 7

GAMES, MAZES, ART, MUSIC, AND ARCHITECTURE

To put it bluntly, Möbius was a bit of a plodder; but when Möbius plodded, he plodded with diligence, elegance, and imagination. He never stopped, and he got places. His great talent was sorting out other people's ideas and seeing them clearly—often more clearly than their creators had done.
—Ian Stewart, "Möbius's Modern Legacy," in Möbius and His Band

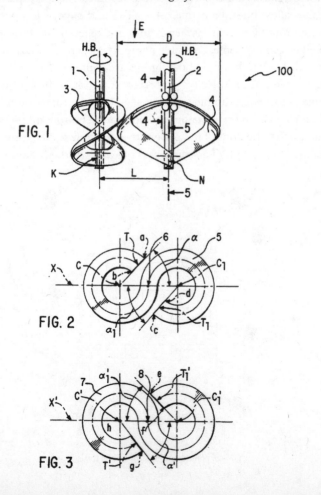

FIG. 1

FIG. 2

FIG. 3

Although many recreational and academic mathematicians are familiar with the role Möbius strips play in topology and are familiar with some of the mathematical properties of these remarkable bands, I find that most people are unaware that this nineteenth-century loop has an interesting role today in literature, art, music, and even games. Various fun and fiendishly difficult games have been devised and played on Möbius strips, Klein bottles, and tori. While writing this book, I've enjoyed playing with tic-tac-toe games, mazes, crossword puzzles, word searches, jigsaw puzzles, and chess—each on tori, Möbius strips, and Klein bottles. The Web contains various computer versions to help players manipulate playing pieces and boards in these strange games. I should point out that playing games on nontraditional surfaces is not just a recreational pastime, but the games allow both mathematical novices and experts to better understand the properties of the surfaces and thus deepen their knowledge.

To whet your appetite, consider figure 7.1, which is a maze played on a torus; the top and bottom of the figures are connected as well as the left and right sides. Your objective is to start at S, travel through the maze, and finish at E. For example, starting at S, you can move up toward the 1 and then continue on the path marked 1 at the bottom of the maze. Try to visualize that the ends of the paper are glued to form the torus and that the path is "connected." The solution is provided in the solutions section.

7.1
Maze played on the surface of a torus.

Figure 7.2 is trickier to visualize because this maze is solved on a Klein bottle, which can be modeled by gluing the left and right sides of the maze together as you would a torus; the top and bottom are glued with a twist. Thus, in this Klein maze, *a* is connected to *a*, *c* to *c*, and *b* to *b*. Your goal is to travel from S to E.

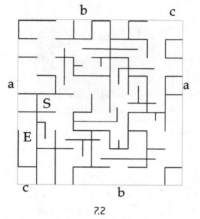

7.2

Maze played on the surface of a Klein bottle.

Möbius Mazes

The first patented Möbius maze that I uncovered is in the form of an intriguing toy invented by David O. McGoveran of Boulder Creek, CA (U.S. Pat. patent 6,595,519, issued 2003). As you can see from figures 7.3 and 7.4, the objective is to get a marble in and out of a hole in a Möbius strip surface. Figure 7.4 shows the channel in which the marble travels. The channel is enclosed by a transparent piece of plastic to hold the marble in the maze. David explains that a puzzle using a Möbius topology and three-dimensional construction makes solving the maze more challenging by "preventing the player from seeing all possible layouts at any one time, as the internal and external surface are both contiguous and identical."

One of the first 3-D Möbius maze toys ever developed and sold on the Internet is the "Moby Maze," designed by M. Oskar van Deventer of the Netherlands (figure 7.5). To solve the Moby Maze, you must push the outer ring along a set of tracks carved into the Möbius strip in order to finally remove the ring from the loop. The solution is not too difficult to uncover; however, playing with this puzzle is very satisfying and graphically illustrates the one-sided nature of the Möbius strip.

Oskar tells me that the Moby Maze was very difficult to build because the three-dimensional modeling program used to design the puzzle

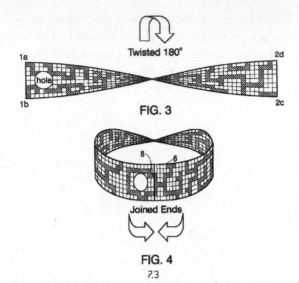

FIG. 3

FIG. 4

7.3
A patented marble maze played on a Möbius strip (U.S. Pat. 6,595,519).

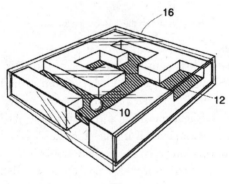

7.4
A close-up of the track in which the marble travels in the Möbius maze
(U.S. Pat. 6,595,519).

became confused with the idea of a three-dimensional object having only one side. Oskar managed to "convince" the program that such an object was possible and worthy of building. The puzzle is available for purchase at George Miller's Puzzle Palace (http://puzzlepalace.com).

To better understand the puzzle, note that the notch at the right side serves both as entrance and exit. The strip has obstacle walls on its surface along both "sides." The topology of the maze includes one 360-degree loop, two dead-end paths, and a long entrance/exit path. This maze may appear simple, but most people get confused by the

7.5

Moby Maze, designed by M. Oskar van Deventer of the Netherlands. To solve the Moby Maze, you must push the outer ring around the Möbius track.

topology of the object and continue traveling round and around, seemingly lost forever. The solution involves making a U-turn at the correct location, which is counterintuitive to most people who try to solve the puzzle. The schematic drawing in figure 7.6 shows the topology of the maze paths without the two dead ends.

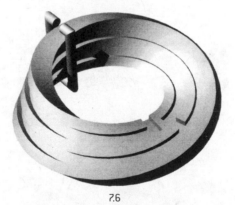

7.6

Schematic figure of the Moby Maze, highlighting the overall topology of the maze.

Chess

When playing a game of chess on a Möbius board like the one shown with the starting configuration in figure 7.7, all kinds of surprises can arise. For example, pieces like the pawn in figure 7.7b can be attacked by pieces on the other "side" of the board. In other words, landing "underneath" a piece captures it as well as landing "on" it. The configuration in figure 7.7c shows how the pawn does not necessarily protect the knight, because the rook may travel in the opposite direction and end up beneath the knight.

If any readers have played Möbius chess, I would be interested in hearing any observations they might have.

Perhaps one of the most fertile grounds for Möbius chess research

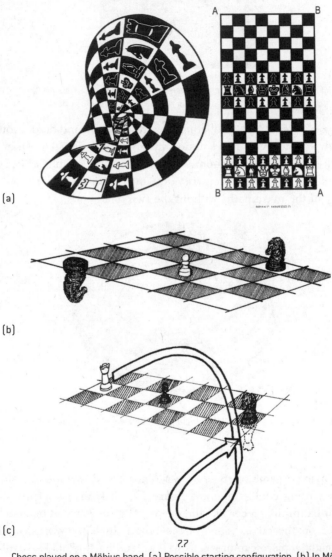

(a)

(b)

(c)

?.?

Chess played on a Möbius band. (a) Possible starting configuration. (b) In Möbius chess, either knight can attack the pawn. (c) In this configuration, the pawn does not necessarily protect the knight because the rook may travel in the opposite direction and end up beneath the knight. (Drawing by Brian Mansfield.)

involves the knight's tour on a Möbius board. An ordinary knight's tour is one in which a chess knight jumps once to every square on the (8 × 8) chessboard in a complete tour. Before discussing Möbius boards, let's review what is known about standard chessboards. The earliest recorded solution of the knight's tour problem is that of Abraham de Moivre (1667–1754), the French mathematician better known for his theorems about complex numbers. Note that in de Moivre's solution (figure 7.8), the knight ends his tour on a square that is not one move away from the starting square. The French mathematician Adrien-Marie Legendre (1752–1833) improved on this by finding a solution in which the first and last squares are a single move apart, so that the tour closes up on itself into a single loop of sixty-four knight's moves (figure 7.9). Such a tour is said to be *reentrant*. Not to be outdone, the Swiss mathematician Leonhard Euler (1707–1783) found a reentrant tour that visits two halves of the board in turn (figure 7.10). (The little squares show positions where the knights transit from one half to the other.)

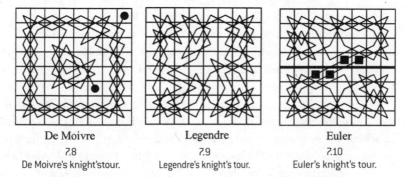

De Moivre	Legendre	Euler
7.8	7.9	7.10
De Moivre's knight's tour.	Legendre's knight's tour.	Euler's knight's tour.

The knight's tour can be created on boards of size five or greater (figure 7.11). The tours shown on the 5 × 5 and 7 × 7 board, are not reentrant. Do you think a computer will ever find a reentrant tour on a huge 2,001 × 2,001 board?

To answer the "2,001 question," consider that a reentrant tour must visit equal numbers of black and white squares. On a 5 × 5 or 7 × 7 board (or any board with an odd number of total squares) a reentrant tour is therefore not possible.

What about knight's tours on Möbius strips and Klein bottles? Professor John Watkins of Colorado College is the leading expert on chess games played on Möbius strips and Klein bottles. In his book *Across the Board*, he theorizes that every rectangular chessboard has the potential for a knight's tour if placed on a Klein bottle.

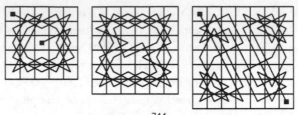

7.11

The knight's tour can be created on boards of size 5 or greater.
Shown here are order 5, 6, and 7 boards.

An m by n chessboard on a Möbius strip with m rows and n columns—
the rows wrapped around the Möbius strip—has a Knight's tour unless
one or more of the following three conditions hold:

   (a) $m = 1$ and $n > 1$; or $n = 1$ and $m = 3$, 4, or 5
   (b) $m = 2$ and n is even, or $m = 4$ and n is odd
   (c) $n = 4$ and $m = 3$

According to Watkins's convention, when a knight moves around the
board and returns upside down on the "other side," this is considered the
same square as the starting square. Because the Möbius strip is a 2-D
surface, we must think of chess pieces as 2-D objects moving inside
the surface of the strip.

Watkins is also fascinated by *domination* of chess pieces on Klein
bottle chessboards. Domination refers to a configuration of chess pieces
in which every vacant square is "under attack." As an example, five
queens are required to dominate an 8 × 8 chessboard, and there are
exactly 4,860 different ways that these five queens can be arranged so as
to dominate the board. There are exactly six ways that two rooks can be
arranged to dominate a 2 × 2 chessboard, and there are 33,514,312 ways
in which eight rooks can dominate an 8 × 8 chessboard.

Figure 7.12 shows how eight kings can be used to dominate a 7 × 7 board
on a Klein bottle. Here, the right-hand side of the board connects to the left
with a twist, and the top and bottom connect without a twist. In general, an
$n \times n$ Klein bottle chessboard can be dominated with $[(1/3) \times (n + 2)]^2 - k$
kings if n is of the form $n = 6k + 1$. This domination occurs on a Klein bottle
with k fewer kings than on a regular chessboard. More generally, the
number of kings required for domination of an $n \times n$ Klein bottle is

   $(1/9)n^2$ for $n = 3k$
   $(1/3)(n+1)^2$ for $n = 3k+2$
   $[(1/3)(n+2)]^2$ for $n = 6k+1$
   $[(1/3)(n+2)]^2 - (1/6)(n+2)$ for $n = 6k+4$

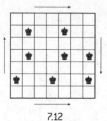

7.12
Kings dominating a 7 × 7 chessboard on a Klein bottle. The arrows indicate the sticky edges of the chessboard. Arrows in reverse directions indicate that corresponding edges are twisted before gluing.

The fact that amateur and professional mathematicians spend their days contemplating chess domination is interesting enough, but when they devote their hours to studying and even playing on Klein-bottle-shaped chessboards, one wonders what else in their lives they enjoy doing in nonstandard ways.

Watkins also tells us that the king's domination number on a Klein bottle for a rectangular $m \times n$ chessboard is given by:

$$y(K_{m \times n}^{klein}) = \left\lceil \frac{m}{6} \right\rceil \cdot \left\lceil \frac{2n}{3} \right\rceil - \left\lceil \frac{n-1}{3} \right\rceil, \; m=1, 2, 3 \bmod 6$$
$$y(K_{m \times n}^{klein}) = \left\lceil \frac{m}{6} \right\rceil \cdot \left\lceil \frac{2n}{3} \right\rceil, \; m=4, 5, 6 \bmod 6$$

The open bracket symbols, $\lceil$ and $\rceil$, represent the ceiling function which rounds up to the nearest integer.

To understand the domination of bishops on a Klein bottle chessboard, examine figure 7.13. Consider a bishop that starts near c on the left side. It moves up to a, goes off the board at a, reappears at the bottom, continues on to b, where it goes off the board again, and then—because of the twist in the Klein bottle—it reappears on the left at b and is now moving down. The minimum number of bishops for domination of an $n \times n$ chessboard on a Klein bottle is given by:

$$y(B_{m \times n}^{klein}) = \left\lceil \frac{1}{2} n \right\rceil$$

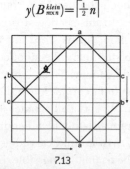

7.13
A bishop's diagonal on a Klein bottle.

Figure 7.14 shows one way in which five bishops can dominate a Klein bottle.

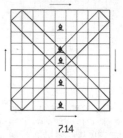

7.14

Five bishops dominate a 9 x 9 chessboard on a Klein bottle. The lines drawn across
the board indicate squares dominated by the second bishop from the top.

Knight's tours on ordinary cylinders are also possible. To visualize this, we can flatten and cut the cylinder so it looks like a rectangle, place "ghost copies" of the basic rectangle at each end, and pretend the corresponding cells are the same as those in the original rectangle. The knight may then move off the edge and onto a ghost, provided it is immediately replaced at the corresponding position of the original rectangle. Tours on a $2 \times n$ cylinder or Möbius band are possible only when n is odd. Tours on a $3 \times n$ and $5 \times n$ cylinder are always possible using a simple repetitive pattern. The height of such a cylinder can be any number of the form $3a+5b$, which includes all numbers except 1, 2, 4, and 7. Even more curious is the fact that several such cylinders can be joined edge-to-edge, and the tours may be combined across the boards by breaking them at suitable places and rejoining them (figure 7.15).

It is known that tours on a 4×4 torus exist. If a tour is possible on an $m \times n$ rectangle arranged in the form of a cylinder, it must also be possible on a torus and a Klein bottle of those dimensions.

Möbius Art Gallery

The Möbius strip has been the basis for countless forms in paintings, etchings, and sculptures. In this section, I present a large international gallery of Möbius and knot forms from artists, designers, mathematicians, and physicists. To start our collection, consider figure 7.16, a contemporary model called "Möbius Stairs," made by British artist Nicky Stephens (www.nickystephens.com). Notice the smooth twists and turns on the railing so that the top surface becomes the bottom and vice versa. Three flights of continuous, laminated handrails, supported on hammered copper spindles, twist around carved ash posts. Stephens says, "I wanted the handrail to be as fluid as possible, inviting the users to follow its twists and turns with their hands."

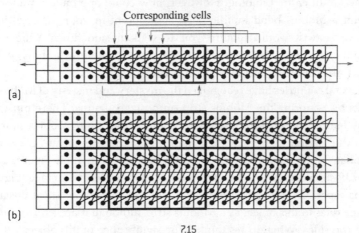

7.15
Knight's tours of cylinders. (a) A 3 x *n* checkerboard on a cylinder. (b) Tours can be visualized by adding "ghosts" on the ends. (c) Tours on several cylinders of different sizes can be joined together by changing the links in a suitable parallelogram.

7.16
Möbius stairs by British artist Nicky Stephens.

Robert J. Krawczyk and Jolly Thulaseedas of the Illinois Institute of Technology in Chicago have considered using the Möbius band as a theme for an entire building. However, how could one build a walkway around a Möbius band building? At some point in your traversal, the walkway's twist would require you to walk upside down! One way around this problem is to create a building that is a hollow Möbius enclosure, with a floor or path suspended within it.

Several countries have recognized the mystery and majesty of Möbius's works by honoring the Möbius strip on postage stamps. There must be many fans of the strip in Brazil; I was able to find three Möbius stamps from this nation. Figure 7.17 is a stamp that commemorates the sixth Brazilian Mathematics Congress in Rio de Janerio in 1967. Figures 7.18 and 7.19 show additional Brazilian stamps of more recent vintage. Figure 7.19 is especially interesting because collectors of mathematical postage stamps refer to this object as a Möbius strip, although it appears to me to have two sides. What do *you* think is the significance of this object?

7.17
Möbius stamp, commemorating the sixth Brazilian Mathematics Congress in Rio de Janerio in 1967.

7.18
Brazilian Möbius stamp.

7.19
Brazilian Möbius stamp.

Figure 7.20 shows a 1969 Netherlands stamp with a Möbius band flattened to a triangle. An almost identical stamp was issued by Belgium at the same time.

7.20
Netherlands stamp with a Möbius band flattened to a triangle.

Figure 7.21 is a Swiss stamp that is part of an annual series of "Europa stamps" that was started in 1957 to emphasize European unity. The stamp series continues today. Each set of stamps has a theme, like "vacation" and "gastronomy." In 1974, the theme was Swiss sculpture, and the design on this particular 1974 stamp represents a sculpture by artist Max Bill (1908–1994). A form similar to the one on the stamp is Bill's 1986 sculpture *Kontinuität* (*Continuity*), which sits outside Deutsche Bank's Frankfurt headquarters.

7.21
Swiss Möbius stamp featuring the work of sculptor Max Gill.

Bill's granite sculpture at the Deutsche Bank is 4 1/2 meters high and is one of his last works. The sculpture depicts the Möbius strip, a motif Bill had explored since the early thirties. Bill took an almost obsessive interest in the Möbius strip, thereby influencing an entire generation of Swiss artists. A huge crane was used to lower this particular eighty-ton sculpture in front of the bank.

Other Möbius strip sculptures decorate buildings and plazas around the world. A stainless steel Möbius strip, eight feet in diameter, casts a tangle of silver reflections in a pool atop Fermilab's Ramsey Auditorium in Batavia, Illinois. A bronze sculpture is installed near an entrance to the Science Center at Harvard University in Cambridge, Massachusetts. Washington, D.C., overflows with beautiful Möbius sculptures. One stainless steel sculpture rests atop a pedestal in front of the National Museum of American History. Another lures visitors to the entrance of the National Air and Space Museum. Even the plaza in front of the U.S. Patent and Trademark Office in Arlington, Virginia, flaunts a Möbius strip made of steel that was painted red. Many of these majestic sculptures are thickened variants in which the "strip's" cross section is essentially an equilateral triangle that is rotated 120 degrees along the strip.

As mentioned in this book's introduction, Dutch artist Maurits Cornelis Escher had a strong penchant for the Möbius strip, which appears in several of his lithographs, including *Möbius Strip I* (wood engraving in four colors, 1961) and *Möbius Strip II (Red Ants)* (wood engraving in three colors, 1963). Even though the pairs of ants in the lithograph seem to be opposite each other, they all exist on the same plane because the Möbius strip has, as we know, only one surface. In *Möbius Strip I*, we see a single bisected band in the form of three fish, each biting the tail of the fish in front. Artist Brian Mansfield has been inspired by Escher's work on Möbius strips and has created his own Möbius forms (figures 7.22 and 7.23). Brian creates numerous Möbius worlds inhabited by robots and other mechanical beings. He is currently working on more complex inhabited, mechanized worlds in the form of Klein bottles, higher-dimensional nonorientable surfaces, and "triply periodic minimal surfaces that have the tetragonal disphenoid as their kaleidoscopic cell" and "Schoen's Manta Surface of Genus 19"–a gorgeous surface that resembles the body of a stingray fish!

In figure 7.23, the Möbius strip allows the robots to travel from one apparent side to the opposite side, representing cycles of creation and destruction, life and death. It is a world in which solenoids and electronic brains may be recycled. According to Mansfield, the robots are self-organizing entities that symbolize the evolution of artificial life-forms that explore endless cycles of metamorphosis. The robots eventually merge into a vast hive mind by the year 2130.

LEGO fanatic Andrew Lipson has created numerous Möbius strips and related knots and surfaces using LEGO pieces. To generate these

7.22
"A Möbius Dr. Möbius" by Brian Mansfield.

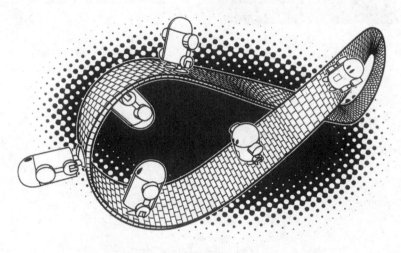

7.23
Möbius strip with robots by Brian Mansfield.

works of art, Andrew writes computer code to guide his creation of the overall shape. He experiments with parameters in the code until he envisions an object that looks attractive and that also has a high probability of being able to balance.

Figure 7.24 is Lipson's LEGO Möbius strip with little men walking on its surface. Figure 7.25 is a LEGO figure eight knot, a knot that we discussed in chapter 2. The figure eight model was among his most difficult sculptures due to the long sweeping curves that hang unsupported in space. Figure 7.26 is a LEGO Klein bottle in which the handle penetrates

7.24
LEGO Möbius strip, © Andrew Lipson.

7.25
LEGO figure eight knot, © Andrew Lipson.

7.26
LEGO Klein bottle, © Andrew Lipson.

7.27
LEGO Klein bottle cross section, © Andrew Lipson.

the main wall of the bottle, as shown in the LEGO cross section in figure 7.27. Lipson's cross-sectiona model actually hinges open so that you can see what he calls the Klein bottle's "digestive tract." He changed the color of the bricks at the top and bottom to emphasize the intersection where the tubes cross. As we have already seen in figure 6.3, each half of the bottle is topologically a Möbius strip.

People on the Web offer all kinds of recipes for creating Möbius strips and clothing. For example, New Jersey computer scientist Mark E. Shoulson describes a way to knit or crochet a Möbius strip with no seams. His site also features him wearing a Möbius strip yarmulke on his head.

Figure 7.28 is a still image from physicist Michael Trott's computer animation showing interlocked gears that turn along the length of a Möbius strip. The gears are arranged in two circles to allow the "first" and the "last" gear to be in sync. Trott holds a Ph.D. in theoretical solid state physics from the Technical University of Ilmenau, Germany, and has been a staff member at Wolfram Research since 1994. He is the author of the four-volume *Mathematica GuideBook for Graphics* and is widely regarded for his encyclopedic knowledge of mathematics and nearly every facet of the Mathematica system.

7.28

Möbius gears, © Michael Trott, reproduced with permission. Adapted from Solution 19c of Michael Trott's *Mathematica GuideBook for Graphics* (Springer, 2004).

Computer programmer and digital sculptor Tom Longtin has also experimented with artistic renditions of Möbius strips involving gears, trefoil knots, and combinations of Möbius strips and trefoil knots. His works are seen in figures 7.29-7.34. Most of these images were created using Tom's own modeling software and rendered using a software package called RenderMan on an SGI computer. Computers indeed provide a powerful means of artistic expression. Tom's Web site, www.sover.net/~tlongtin/, has additional examples.

Although figure 7.29 appears to be rather complex, it still retains the

Möbius strip character. If one takes a strip of paper, twists one end 180 degrees relative to the other (a half turn) and glues the ends together, then all the teeth profiles in this figure could be drawn upon the surface and the spaces cut out. Figure 7.30 represents a strip of paper that has been twisted 540 degrees (three half twists) before being formed into a knot and having its ends glued together. Once Tom creates this basic motif, he cuts holes through the strip, which retains the original topology of both a trefoil knot

7.29
Möbius gears by Tom Longtin.

7.30
Möbius and trefoil knot with gears by Tom Longtin.

7.31
Möbius and trefoil knot puzzle by Tom Longtin.

7.32
Möbius puzzle by Tom Longtin.

7.33
Trefoil puzzle by Tom Longtin.

7.34
Möbius-like object with holes by Tom Longtin and Rinus Roelofs.

and a Möbius strip. Figure 7.31 is made by giving a strip of paper three half twists and then forming a knot by connecting the ends. This, too, is both a Möbius strip and trefoil knot. Figure 7.32 is an exploded view of puzzle pieces showing how they would fit together in a Möbius strip. Figure 7.33 is a classic trefoil knot with hexagonal puzzle shapes drawn upon the surface. Figure 7.34 shows a Möbius strip with holes. In this peculiar arrangement, a Möbius strip is wrapped onto itself. In an ordinary paper Möbius strip, we would travel once around while twisting 180 degrees. This one requires two trips around to twist 180 degrees. Like a traditional Möbius strip, this form had one side and one edge before it was punched with holes.

7.35 Möbius strip and knot by
Rob Scharein.

7.36 Möbius strip and knot by
Rob Scharein.

7.37
Möbius strip and knot by Rob Scharein.

Rob Scharein, a researcher who develops educational software for visualization in mathematics and science, combines his love of art and mathematics when creating the knotted and linked Möbius strips in figures 7.35–7.37.

All of the ribbons in these plots are Möbius-like (i.e., nonorientable surfaces). Scharein uses his custom-designed software KnotPlot to produce these plots, and he encourages readers to experiment on their own with his software, which you can download for free from the KnotPlot Web site (www.pims.math.ca/knotplot/). Among other things, he uses his software to check that the strips are all nonorientable, as he doesn't want to verify by eye some of his more complex knots! Rob is also one of the world's leading experts on visualizing extremely complex knots, such as those shown in 7.38 and 7.39.

7.38
Complicated knot by Rob Scharein.

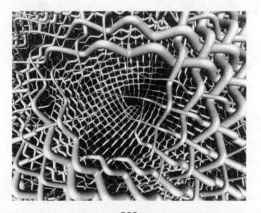

7.39
Complicated knot by Rob Scharein.

Teja Krasek, a well-known Slovenian artist, spends her time creating Möbius strip sculptures adorned with Penrose tiles (figure 7.40). This pattern of tiles, discovered by English mathematical physicist Roger Penrose, can completely cover an infinite surface, but only in a pattern that is *aperiodic* (nonrepeating). In other words, the tiling pattern does not repeat periodically like the hexagonal tile patterns on some bathroom floors. When tiling the Möbius band, Teja uses two different tiles shapes, each having four sides of the same length. In particular, one rhombus tile has four corners with the angles {72, 72, 108, 108} degrees, and the other has angles of {36, 36, 144, 144} degrees. When forming the Penrose tiling, no two tiles can touch so as to form a single parallelogram. Given this restriction, an infinite number of ways exist to tile an infinite plane and still leave no gaps in the tiling. The resultant pattern will always be aperiodic so that the pattern never repeats exactly. Scientists are aware of numerous real-world quasicrystals whose atoms are arranged in the same pattern as a Penrose tiling.

7.40
Penrose tiling on Möbius strip. Sculpture by Teja Krasek.

Naturally, the challenges of forming a Penrose tiling on a Möbius strip are many for Teja. For example, she must ensure that tiles perfectly join as the two "ends" of the strip meet in the final, single-sided object. Additionally, Teja designed the tiling so that the apparent triangular segments that touch the edges of her sculptures would form the appropriate rhombi if the two edges were attached. Yet another challenge involves her coloring the Penrose Möbius strip using only three colors. In 2000, mathematicians Thomas Sibley and Stan Wagon proved that a planar configuration of such tiles can be colored using only three colors in such a way that adjacent tiles receive different colors.

When Teja creates these sculptures, she starts by drawing or printing patterns on paper in both their original and mirror forms. When the tiles are finally glued to the strip, she must stick the same tile on both "sides" of the strip so that the colored tiles on one side occupy the same position and have the same color as on the "reverse" side. She is currently working with translucent materials that enable a single tile to be viewed on either side, which saves her both work and her sanity. Additional examples can be found at her Web site, http://tejakrasek.tripod.com.

I should add that Teja's Christmas tree is always decorated with the most beautiful, shiny, silver and gold Möbius strips I have ever seen (figure 7.41). The strips glisten with sparkling stars along their surfaces, and the tree is enough to warm the heart of any romantic mathematician. Teja reminds me that we do not need any strings to secure Möbius ornaments to the tree because they hang on the branches through their centers.

7.41
Silver and gold Möbius Christmas tree ornaments by Teja Krasek.

For a 2005 snow-sculpting competition held in Breckenridge, Colorado, a team of sculptors rendered a split, triply twisted Möbius strip designed by computer scientist Carlo H. Séquin of the University of California at Berkeley (figure 7.42). In addition to Séquin, the snow-carving team consisted of mathematicians Stan Wagon of Macalester College in

St. Paul, Minnesota, John Sullivan of the Technical University of Berlin, Dan Schwalbe of Minneapolis, and Richard Seeley of Silverthorne, Colorado. The sculptors started with a 10' × 10' × 12' snow block and spent the first two days simply removing half of the twenty tons of snow in their block to obtain a rough approximation of a triply twisted band.

7.42
"Knot-Divided," snow sculpture by Team Minnesota, Breckenridge, Colorado, 2005.
(Design: Carlo H. Séquin, U.C. Berkeley; photo by Richard Seeley)

Möbius Music

As we discussed in chapter 6, if you were to travel within a Möbius universe, you would return to your starting point with your left and right sides reversed. Travel around the strip again, and you'll return to your starting point with your organs back to their standard orientation. Similarly, Möbius music can be created by pasting a musical score to a Möbius strip. The music is played as usual the first time around. When the musician has arrived at the starting point, the music is played again but with some geometrical variation. For example, the second time around the score may be mirrored or played upside down.

Johann Sebastian Bach wrote Möbius-like music such as his *Crab Canon*, in which the musician can play from start to finish and then flip the musical score upside down and play it again. Austro-Hungarian composer Arnold Schoenberg, several centuries later, experimented with crab canons, which he called "mirror canons."

Although Schoenberg was a musical genius from an early age, some of his more unusual works were not well received. When his Chamber Symphony no. 1 was played in a 1913 concert, the audience booed. Later in the concert, during a performance of some songs by Austrian composer Alban Berg, fighting broke out, and the police were called to keep the peace.

Schoenberg, who was an excellent painter, was also superstitious and feared the number thirteen. In fact, he titled one of his operas *Moses and Aron* rather than *Moses and Aaron*, deleting an *a* because the correct spelling had thirteen letters.

Russian-American composer and linguist Nicolas Slonimsky was directly inspired by the Möbius strip. Two singers and one piano player first performed his "Möbius Striptease" in 1965 in Los Angeles. Here are some of the lyrics from the piece:

> Ach! Professor Möbius, glörious Möbius
> Ach, we love your topological,
> And, ach, so logical strip!
> One-sided inside and two-sided outside!
> Ach! Euphörius, glörius Möbius striptease!

The instructions on the score read, "Copy the music for each performer on a strip of 110-b card stock, 68" by 6". Give the strip a half twist to turn it into a Möbius strip." The song was, in essence, a perpetual vocal canon written on a Möbius band to be revolved around the singers' heads during the performance.

Nicolas Slonimsky (1894–1995) came from a long line of Jewish intellectuals on his mother's and father's side. His relatives and forebears included novelists, poets, literary critics, university professors, translators, chessmasters, economists, mathematicians, inventors of useless artificial languages, Hebrew scholars, and philosophers. Slonimsky always had big ambitions and as a teenager wrote his own future biography, in which he speculated (inaccurately) that he would die in 1967.

In 1945, Slonimsky became a lecturer in Slavonic languages and literatures at Harvard University. His musical compositions focused on odd structures, and some songs were set to text from tombstones. His orchestral work *My Toy Balloon* (1942) was a variation on a Brazilian song, the score of which included the instruction that one hundred colored balloons be exploded at the climax. He was also famous for his "grandmother chord" containing twelve different tones and eleven different intervals.

Today, several musical groups have "Möbius" in their names. The Möbius Band from Massachusetts is a contemporary musical trio that uses traditional instruments (guitar, bass, drums, and voice) and modern electronic ones (synthesizer, sampler, and electronic percussion). The Möbius Band should not be confused with Möbius Donut, an Oakland, California, musical group heavy on melody and groove. Korean musician Jo Yun from Jaeju Island used multiple synthesizers and an acoustic guitar to produce his CD titled *Möbius Strip*. The album opens with the clanging of church bells, which morph into a tribal rhythm with drums. The back cover of the album has four separate flaps, each having a picture of a peacock feather.

Musician Peter Hammill's song "The Möbius Loop" has lyrics such as "Indecision and uncertainty catch you now. . . . How you're gonna take sides now you're on the Möbius loop?" Infinity Minus One, a hard rock and metal band from Boston, recorded their first CD, *Tales from the Möbius Strip* in 2002. Their music has diverse influences, including rock, metal, film scores, and video games.

◉ Cutting Devil Configurations

Figure 7.43 shows three paper-strip constructions involving twisted "arms" and a hole. What do you think happens when you cut around the center hole of these figures along the dotted line? The first configuration has one twisted connector, the second has two half twists in the same direction, and the third diagram shows an object that has two half twists in the opposite direction. To help visualize the configurations, try to create the strips with paper and actually perform the experiments. The easiest way to create the models is to cut two oval regions as shown in figure 7.44. The dashed lines are the guidelines for cutting. To form the closed loop, simply tape the ends together with the desired number of half twists.

Can you predict what will happen if you cut along the dotted lines in the two

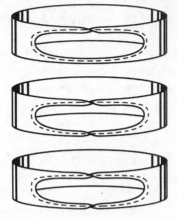

7.43

What happens when you cut around the center hole along the dotted line?

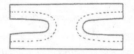

7.44 Constructing the devil configurations.

shapes in figure 7.45? Here, we construct two loops of the same length and width. In one configuration, one of the arms has a twist. You can create these figures by cutting a piece of paper into an X shape and then gluing the arms together.

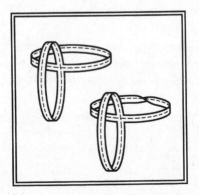

7.45 What happens when you cut along the dotted lines?

Möbius Strip in Psychology and Human Relations

❧ *Memory for the survivor, he said, is like a Möbius strip. Past, present and future are connected and the experiences situated anywhere on the loop are accessible. In therapy, we have the opportunity to ride that loop, touch past experience and relate it to the present. In other words, we can be topographers of our own lives.*
— Marjorie Levenson, "The Möbius Strip"

❧ *Langdon smiled. "You must be a teacher too."*
"No, but I learned from a master. My father could argue two sides of a Möbius strip."
Langdon laughed, picturing the artful crafting of a Möbius strip—a twisted ring of paper, which technically possessed only one side.
— Dan Brown, Angels and Demons

❧ *With striking imagery, Rilke offers us a mystic's map of wholesomeness, where inner and outer reality flow seamlessly into each other, like the ever-merging surfaces of a Möbius strip, endlessly co-creating us and the world we inhabit.*
— Parker J. Palmer, The Courage to Teach: Exploring the
Inner Landscape of a Teacher's Life

❧ *Freud's logic was a veritable Möbius strip of circularity. When patients complied with his insistence that they remember early sexual material, he called them astute; when they did not, he said they were resisting and repressing the truth.*
— Thomas Lewis, Fari Amini, and Richard Lannon, A General Theory of Love

❧ *As the scenes—and lovers—play against each other, hope clashes with sorrow, ambition rings against frustration, a marriage is dashed on the rocks and pieced back together only to be broken again. The effect is like twisting a wedding ring into a Möbius strip.*
— Chris Page, "Clever Device, Not a Moving Story, Fuels 'The Last Five Years,' "
Get Out 2005

CHAPTER 8

LITERATURE AND MOVIES

*[Möbius had] no body of deep theorems . . . but a style of thinking, a working
philosophy for doing mathematics effectively and concentrating on what's
important. That is Möbius's modern legacy. We couldn't ask for more.*
—*Ian Stewart, "Möbius's Modern Legacy," in* Möbius and His Band

*When a man and woman join as lovers, there is a potential infinity of rela-
tionships that, like the Möbius strip, has no beginning and no end . . .*
—*Carol Berge,* A Couple Called Möbius:
Eleven Sensual Short Stories

"Möbius strippers never show you their backside."
—Joke circulating on the Internet

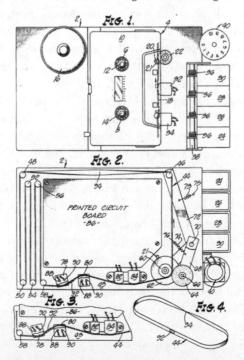

Möbius Stories: The Literature of Nonorientable Surfaces

So many stories exist in which the Möbius strip plays an important role that the following is merely a sampling of some Möbius references in literature and movies. Stories focusing on the Möbius strip had a heyday in the 1940s, so we will start our journey there.

One of the earliest and most creative short stories on the Möbius strip is Martin Gardner's "No-Sided Professor" (1946), which appeared in Clifton Fadiman's *Fantasia Mathematica*. In the story, members of the Möbius Society—an organization of mathematicians working in the field of topology—meet with a Dr. Stanislaw Slapenarski. As they gather around a dinner table, replete with silver-plated napkin rings shaped like Möbius strips and Klein bottle coffee mugs, Dr. Slapenarski explains his breathtaking topological discovery.

Dr. Slapenarski's lecture begins with his uncovering of August Möbius's "lesser known treatise" on how to turn an ordinary loop with two sides into a Möbius strip with one side. In this (mythical) treatise, Möbius says that there was no theoretical reason why a surface could not lose *both* its sides to become a no-sided surface!

The professor stares at his rapt audience and explains that the no-sided surface is difficult to imagine, but that doesn't mean it is not real or practical. Many concepts in mathematics are inconceivable, including higher-dimensional geometry—but that is "no basis for denying either their validity or usefulness in mathematics and modern physics."

Moreover, even a one-sided surface is inconceivable to anyone who has not seen and manipulated a Möbius strip. The professor explains that people who are handed a Möbius strip to play with sometimes are still unable to understand how it has just one side. Given this, the fact that we cannot imagine an object does not mean it cannot exist.

The professor then proceeds to fold a piece of paper into a no-sided "Slapenarski surface," using an intricate procedure involving scissors, paste, and pale blue paper. At the end of his folding sequence, he smiles at the audience and presses one of the projecting ends of the paper against the other, and the paper figure vanishes in his hands! It has become a zero-sided surface. When the mathematicians in the room think this is nothing more than a parlor trick, Slapenarski becomes angry and forcefully folds one of the mathematicians into a no-sided surface by manipulating the man's arms and legs. The mathematician disappears, leaving only his clothes behind. The audience gasps and chaos ensues.

In Arthur C. Clarke's 1946 short story "The Wall of Darkness," the protagonists live in a universe consisting of only one star and one planet named Trilorne. A mysterious, impenetrable wall surrounds the entire habitable region of Trilorne, a world in which all exploration is prevented by the wall that appears to extend to the heavens. Civilizations on Trilorne have always wondered what is on the other side of the wall. Some Trilorne philosophers say, "What is beyond, we shall discover when we die, as that is where the dead go." Others say, "Behind the Wall is the land where we lived before we were born. If we could remember that far back, we would know the answers." A few wise people worry that the wall was built to keep something dangerous from entering their world.

Finally, a rich man and his engineer friend determine a way to scale the wall by building a great stairway along its edge. Their arduous mission is to determine what is on the other side. At the end of the quest, they learn that they are living on a Möbius strip and that by going over the wall, they merely enter their world from the other side.

For reasons that I don't understand, this discovery of what is on the other side of the wall is so objectionable that the two explorers decide to blow up the stairway so that no one else can learn the secret of their world. In effect, the purpose of the wall is to prevent the world's inhabitants from making the complete trip around the strip to learn of the strange topology of their space. Perhaps the wall is useful because it gives the inhabitants a sense of mystery, prevents them from traversing paths that reverse their orientation and handedness, or prevents the discovery of new routes for waging war. Clarke never reveals why the protagonists decide to destroy the great stairway and keep the shape of the world a secret.

In William Hazlett Upson's "A. Botts and the Möbius Strip" (1945), a Möbius band actually saves the lives of several Australian soldiers. The story takes place in the year 1945, when Major Alexander Botts needs a way to distract the uncooperative Lieutenant Dixon. He finally decides to occupy Dixon's time by having him paint a belt that runs through two holes in a pump house wall. Secretly, Botts unlaces the belt, gives it a half twist, and laces it together again to form a Möbius strip. When Dixon tries to paint the outside of the belt, as he is instructed, without painting the inside, he becomes so confused, delayed, and enraged that Botts has plenty of time to abscond with a tractor desperately needed for the survival of Australian soldiers in New Guinea.

In the same author's "Paul Bunyan versus the Conveyor Belt" (1949), uranium miners use a mile-long conveyor belt in the shape of a Möbius strip to transport ore. The protagonists in the story argue at length about what would happen if the belt needed to be cut in order to make it longer. As the mine lengthens, Bunyan decides to cut the belt down the middle to increase its length.

"That will give us two belts," said Ford Fordsen. "We'll have to cut them in two crosswise and splice them together. That means I'll have to go to town and buy the materials for two splices."

"No," said Paul. "This belt has a half twist—which makes it what is known in geometry as a Möbius strip."

The miners renew their arguing when they need to lengthen the belt again, and wonder about the results of cutting the lengthened strip.

When A. J. Deutsch wrote "A Subway Named Möbius" in 1950, he was a member of the Harvard astronomy department. He was probably getting tired of the traffic while commuting to work when he wrote this story of the Boston subway system, which becomes so complicated and looping that it finally forms a Möbius strip that spans dimensions! Part of the subway remains in our world, while one loop goes into a higher dimension. Trains make clattering noises, seemingly nearby, but cannot be seen. When attempting to explain it, one of the characters in the story says that a new piece of track "has made the connectivity of the whole subway system of an order so high that I don't know how to calculate it. I suspect the connectivity has become infinite."

The 1996 movie *Möbius*, directed by Gustavo Mosquera, features a train in the Buenos Aires subway system that suddenly vanishes. The plot has many similarities to "A Subway Named Möbius." Because the subway system has had so many additions and has grown so vast, nobody is able to picture it anymore, not even the train engineers. One day, a train disappears, and people can hear the train rushing through tracks, but can never seem to find it.

The subway manager tries to come up with an explanation for this phenomenon and asks the engineer responsible for the growing subway complexity to come talk to him. The engineer resists and sends Daniel, a mathematician friend, to the manager to help with the investigation.

Daniel attempts to obtain the subway layout plans from a mysterious Dr. Mistein, who, alas, is not home and cannot be located. Daniel contemplates the problem and comes to believe that the subway system, with its countless additions over the years, has become so complex that a gigantic Möbius strip has unintentionally formed, and the missing train is now trapped on the loop. The subway manager scoffs at the idea of the Möbius strip but decides to shut the subway system down in an effort to avoid further disappearances.

Even though Daniel's theories are not seriously considered, he continues his investigation of the subway. The majority of the movie takes place in subway tunnels through which Daniel travels in order to understand the subway layout. One night when he boards a subway on his way home, he discovers that he is aboard the lost train! He walks to the first car of the train and finds that the missing Dr. Mistein is driving it.

Although the idea of a disappearing subway train first came from the story "A Subway Called Möbius," Mosquera conceives the idea of the missing train as a metaphor for the people who disappeared during the dictatorship periods in Argentina. Mosquera says that his engineering studies in college helped him "appreciate mathematics and abstract ideas and the artistic works of people like M. C. Escher [so that the concepts] began to all come together" in his film. Indeed, the film features a mathematician hero—rare for movies these days—and several references to advanced geometry concepts. Mosquera employed forty-five students to help find suitable locations for filming; one such location was an abandoned Buenos Aires subway station.

The Möbius strip was also referred to in "Time Squared," an episode of *Star Trek: The Next Generation*. The starship USS *Enterprise* encounters a mute and agitated Captain Picard from six hours in the future. The present Picard worries that whatever judgment he made in the future must have left him and his crew in a never-ending cycle in which an old *Enterprise* keeps rediscovering a Picard from the future. In the episode, Lieutenant Worf remarks, "There is the theory of the Möbius, a twist in the fabric of space where time becomes a loop from which there is no escape." Geordi responds, "So, when we reach that moment—whatever happened will happen again . . . The *Enterprise* will be destroyed, the 'other Picard' sent back to meet with us and do it all over again. That sounds like someone's definition of hell."

Several stories written for children or young adults incorporate the Möbius strip in their plots. Amy Cameron's *The Secret Life of Amanda K. Woods* (1998) features a Möbius strip on the cover. The main character, eleven-year-old Amanda from Wisconsin, is a whiz at mathematics. One day, a friend's mathematician parents give Amy a Möbius strip to examine. She immediately understands that it is one-sided.

Amy is told, "It is called a Möbius strip. It is important to geometry. And in life, too, sometimes the outside turns into the inside and the inside into the outside." The Möbius strip becomes Amanda's metaphor for wisdom, growing maturity, and ability to manage opposing demands.

Mark Kashino's book *The Journey of Möbius and Sidh* (2002) includes a three-foot-long Möbius strip printed with the story's highlights. The strip is laminated for repeated use, and the book also includes an erasable marker. The publisher says, "The peculiar properties of the Möbius Strip seem an unusually appropriate metaphor for our lifelong search. The characters are non-ethnic and multi-colored."

Another creative biological use of a Möbius strip in science fiction occurred in my novel *The Lobotomy Club* (2002). In the book, a brain surgeon named Adam discovers that a certain Möbius topology of neurons in the brain creates a portal to new realities. Here is a snippet of dialogue between Adam and a beautiful woman named Sayori:

Adam closed his eyes. "Why am I here?"

Sayori was now petting the cat, which stretched out beside her and purred. "I know about your work on the CMS—the Cerebral Möbius Strip." Her eyes seemed to blink whenever the cat's did.

Kierkegaard earnestly searched for something in a used Chinese-food container, and then tossed the box into the trash. He settled for a hexagonal pill the color of seaweed.

Wasabi looked questioningly from Adam to Sayori. "CMS?"

Sayori nodded. "The CMS is a special topology and network of neurons that Dr. Wolf discovered residing in several priests' brains after they had ecstatic visions, convulsed, and died a day later. Two Tibetan monks reported the same kinds of visions and also died."

Ikura stopped chewing her gum. "Why did the CMS form in these people?"

Sayori rubbed the cat's pillowy belly. "We don't know," she said. "We do know that it allowed them to experience transcendent feelings and to perceive reality in heightened ways. Adam nicknamed this rewiring the 'Cerebral Möbius Strip' because the neurons doubled back on themselves in a figure eight."

The characters in the book learn that our baseline reality is an illusion, and the CMS can help them experience what might be a truer reality. Adam agrees to help members of the Lobotomy Club induce the CMS in their brains so that they can safely peer into new worlds.

My favorite Möbius-shaped animal in literature is the cow named Moobius in Ian Stewart's *Flatterland* (2001). Moobius is intelligent and has an extraordinarily long tail that wraps all the way round to touch his face. The tail is glued to his nose. Moobius explains that he has two sides *locally*–but viewed as a whole–the twist in his tail makes the two sides become one.

Perhaps the sexiest book with Möbius in the title is Bana Witt's *Möbius Stripper* (1992), which describes a woman's adventures in the sexual and drug underground of San Francisco during the 1970s. The book opens with the nineteen-year-old narrator contemplating the possibility of acting in porno films. The plot includes a fascinating collection of short snippets derived from the author's life, which includes sexual and drug-induced experiences. The book is quite lively and not for the prudish.

Möbius-Structured Literature

The Möbius strip not only appears in movies and literature, but it has been used as a model for strangely looping plots. In Möbius-structured literature, the plot is sometimes recursive, an echo of itself, or characters return to the beginning of the story in a slightly altered form–as in Frank Capra's *It's a Wonderful Life* (1946), in which George Bailey has the option of returning to an earlier time in his life with new wisdom.

Of course, this is not literally a Möbius strip in the mathematical sense, but many have used the metaphor of the Möbius strip to describe these odd plot circuits, which are often quite mysterious and emotionally moving. For example, science fiction writer Samuel R. Delany's 800-page novel *Dhalgren* is full of Möbius-like allusions. One of the main characters, Kidd, writes a book that might be the actual text of *Dhalgren*. Every now and then, the flow of time seems to stop. Kidd walks in one direction and ends up in another direction. Building locations shift. Days pass in the blink of an eye, or in some locations, seconds last for hours. The final chapter focuses on a notebook that Kidd finds. Kidd writes in its margins, and it seems he has written in the margins before he discovered the notebook. In the end, the notebook consumes itself and the world destructs. The book finishes on a sentence fragment that leads back to a plot very similar to the beginning of the book, as if the plot were stretched out on a Möbius loop with the end mirroring the beginning, with character roles reversed.

Marcel Proust's *In Search of Lost Time* (1913) also contains major and minor Möbius loops as the main character Marcel returns to his past to reflect on his life. Sometimes, time seems to disappear entirely from Proust's work. We spend hundreds of pages examining the nature and ideas of a character or a situation, while there is minimal flow of time. In "Proust's Ruined Mirror," Jonathan Wallace writes, "In Proust's novel, time is a river in which the characters swim; it *tends* to carry them downstream, but like fish, they occasionally reverse themselves and struggle against its flow." Proust's greatest desire was to travel through time, to recapture the past with its lost memories and people. In some ways, *In Search of Lost Time* resembles a chunk of spacetime that contains past, present, and future. In this chunk, the reader and Proust may explore the story like they would a hyperspace palace, wandering in time and space through rooms anchored in different epochs.

Proust's work also focuses on various physical paths through town that suggest a Möbius strip. In particular, the character of Marcel reminisces about his early years spent with relatives in the town of Combray. At one end of his aunt's house is a door that leads to a walking path called Meseglise Way, also called Swann's Way. The other leads to Guermantes Way. On one level, they are just paths that traverse the village and on which Marcel's family takes daily walks. One path goes to the estate of the wealthy Guermantes family, the other to Swann's middle-class estate. However, they represented much more to Proust—different directions in life and the choices we make. At the end of his masterpiece, the narrator, who has grown old, revisits Combray and discovers a shortcut that unites the two paths. He realizes now that the two "ways" are connected after all.

> Thus for me, do the Meseglise Way and the Guermantes Way remain linked to so many small events of that one life of all the diverse lives that we lead on parallel lines, the one which is the fullest of events, the most rich in episodes, the life of the mind.

Although the Guermantes Way leads to the elegant château of the aristocratic Guermantes family, Proust never actually seems to reach the château because the walking distance is too great. Thus, one path represents a path to the ordinary, and the other represents a path to the furthest reaches of space, time, and mind. I delve into Proust's work in greater detail in my book *Sex, Drugs, Einstein, and Elves.*

The comedy *Six Characters in Search of an Author* (1921) by Sicilian-born writer Luigi Pirandello (1867–1936) also has a wonderful Möbius plot. The protagonists of the comedy are six characters who have been created by their author but left in an unfinished drama. They arrive at a rehearsal of a Pirandello play and convince the director to allow them to perform their drama for him so that they can become whole characters. The director eventually agrees to become an author for their new lives. During the course of the play with these six characters, some of the characters die, and the director cannot tell if they are acting or actually dead. In the end, neither he nor his actors are able to tell what is real.

In 1937, British writer John Boynton Priestley (1894–1984) presented *Time and the Conways*, a play in which the action at the end of the second act is thirty years later than in the first act, and then in the third act, the play loops back to the end of the first act. Thus, in some ways the third act might be considered a misplaced middle act. The play begins in 1919, when the affluent Conways are joyfully celebrating Kay's twenty-first birthday. The scene jumps to 1938, when the family is again assembled, but Europe is on the edge of war. Finally, we return to 1919, and our advanced knowledge gives a strange dramatic irony to the events that unfold. At a deeper level, the play makes the audience wonder whether true happiness is possible, whether or not we can change our destinies, and it reinforces an idea that time is not linear and that the past and future are always present with us.

The movie *Donnie Darko* (2001), directed by Richard Kelly, is a blend of supernatural thriller and time travel paradox that focuses on sixteen-year-old Donnie who lives in suburban Middlesex, Virginia. A demon tells him that the world will end in twenty-eight days, sixteen hours, forty-two minutes, and twelve seconds. Throughout the movie, Donnie sees liquidlike tubes protruding from people's bellies and pointing in the direction that person will move in the near future. Donnie Darko sees his own lifeline stretching from his belly, as if his actions have been predetermined, and he's a pawn, trapped in the jejune jardinière of time.

The plot has a strangely looping story that leaves most moviegoers bewildered and discussing the movie for weeks. In the end, the movie returns to its opening scene, but this time Donnie has foreknowledge and is presumably able to save those he loves by sacrificing himself. Film critic Jim Emerson, editor of RogerEbert.com, says that the film's opening with Donnie waking up on a hillside road at dawn is "essential to the movie's endlessly circular (or Möbius-strip) form, and part of what draws you back again. It begins with a scene that belongs at the end of the last time you watched it—a dream within a dream within a dream . . .

And when you think about it that way, it helps locate the entire movie in the space-time warp between Donnie's ears." I enjoyed the movie. See it and enter a movie form of the Möbius strip.

Many other movies and stories include a time loop in which characters return to an earlier time in the movie with the ability to relive the past with greater knowledge and to remake their lives. In Brian De Palma's 2003 movie *Femme Fatale,* Laure Ash is a thief who has the mysterious and unexplained opportunity to live the movie again and choose a wiser path through life. In my book *Liquid Earth,* the character Max has the opportunity to live the entire book again, and renders hope that with his new knowledge, he will be able to save the world from reality fractures.

In *50 First Dates* (2004), Lucy Whitmore undergos endless successions of Möbius-strip lives, as she wakes up each morning with no memory of having met Henry Roth the day before. Lucy is afflicted with short-term memory loss after a car accident, and she's caught in a perpetual loop. To her, every day is the same Sunday in October, which of course makes it nearly impossible to form new relationships. Henry falls in love with her and tries to imagine ways in which a deep relationship is possible. Gradually, despite her handicap, some small strand of her mind seems to find its way into the next Möbius strip day, until she finds herself painting her lover's portrait, even though she cannot remember who he is.

In my book *Time: A Traveler's Guide,* I give surprising Möbius scenarios that involve time travel paradoxes and causal loops. Let's consider one of my favorite plotlines that will surely twist your mind. Figure 8.1 schematically represents the characters' paths through space and time. (Assume that the characters have a time travel machine.) In this figure, I represent myself by the ♂ in the center, and Monica, the woman I love, by the ♀. Let's assume we initially meet at the position in spacetime marked by the 1. A little later, at the position marked by the 2, we marry and have a baby daughter, Monica Jr. Her path through life is represented by the dashed line. Unfortunately, Monica Jr. is abducted by a stranger at birth, and we never see her again. She grows up, and at age twenty (marked by 3) she decides to go back in time to find her roots. After traveling back in time, she spends twenty years growing up and having a fairly normal life. Finally, she meets me at 1! We fall in love, marry, and the rest is, as they say, history. She is the woman I initially met at 1. Meanwhile, at the position marked 4, the "original" Monica Sr. and I decide to go back in time in hopes of finding our lost daughter. We go back in time, and at 5 we have a baby boy who grows up (wiggly line in figure) to be me 1. At the very bottom of the figure, the "original" Monica and I go way back and visit prehistoric cavemen. Notice that Monica is her own mother and grandmother, and I am my own father and grandfather.

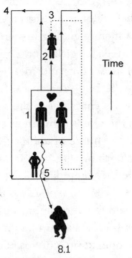

8.1

If time travel is possible, then world-lines might become closed loops. I meet Monica (1), and have a baby daughter, Monica junior (2), represented by the dashed line, who grows up (3) and decides to travel back in time. Monica junior grows up and meets me at (1)! See the text for all the details.

This scenario does seem quite crazy. After all, who is Monica's mother, father, grandfather, grandmother, son, daughter, granddaughter, and grandson? Monica Jr. and Sr. are the same person. If we draw more of Monica's family tree, we might find that all the branches are curled inward and back on themselves, as in a loop. She can be an entire family tree unto herself. This is an example of a paradox unlike the one where a person goes back in time and kills his grandmother, thus altering the past. In the case illustrated in figure 8.1, characters are *fulfilling* the past, not destroying it. Thus the lines in the schematic representation (called world-lines by physicists) travel in a *closed loop*, fulfilling rather than changing the past.

Another Möbius plot occurs in Gabriel Josipovici's stories collected in *Möbius the Stripper* (1974), which deal with a man who is nervous about his writer's block. Möbius's story is displayed on the top of each page, and the text of the narrator's story about Möbius is in the lower half of each page. At the end of the top story, Möbius kills himself, which creates a stark blank page that confronts the narrator in the bottom half story. Toward the end of the narrator's story at the bottom, he finally overcomes his writer's block and starts to write Möbius's story printed at the top.

In a similar vein, *The Gift* (1937) by Vladimir Nabokov features a protagonist named Fyodor. Fyodor is a Russian living in Berlin, and he is having a great deal of difficulty getting his writings published. Near the end of the book, Fyodor tells his girlfriend Zinia that he wants to write a book about how he started writing and met her. It seems that the book Fydor wants to write is the book the reader has been reading! In this sense, Fyodor is no longer a character in the novel, but its author.

In Möbius literature, the plot is sometimes recursive, an echo of itself, or one plot exists within the frame of another. I've heard the term "metalepsis" sometimes used when referring to times in Möbius plots in which the characters cross boundaries between layered plots. For example, in Coleman Dowell's novel *Island People* (1976), a low level becomes the top level, taking over the narrative and creating a kind of Möbius band. The story involves an unnamed man who leaves the city to live in a house he has bought on a tiny island. The man appears to be a loner or an outsider among the "island people," who live on the island year-round. Though he lives a solitary life with his dog, he does enjoy occasional visitors from the city. Suddenly, the reader realizes that this tale of the loner on the island is the story "The Keepsake," written by another unnamed man living under circumstances identical to those of the first man, though somewhat more isolated from the world beyond his island. Reviewer Christopher Sorrentino, writing for *Center of Book Culture*, explains, "It's a book that doesn't seem to have been written as much as it seems to crawl out of itself . . . [The book's character avatars] echo one another across the chasm of the novel . . . Countless parts of *Island People* set off sympathetic vibrations with countless other parts." Eventually, the man invents a female alter-ego, who haunts him as his mind disintegrates.

In Daniel Hayes's *Tearjerker* (2004), we encounter Evan Ulmer, a frustrated writer discouraged by his growing collection of book rejections but eager to learn more about the book business. He kidnaps an editor from a prestigious New York publishing house so that the editor will explain the process to him. It turns out that Evan has written a book about a failed writer kidnapping an editor, and he would like to get this book published. During the week that Evan kidnaps his victim, he also meets a strange woman named Promise who uses Evan as a character in a novel she is writing. In it, he's having an affair with a fifty-year-old woman. She wants Evan to meet her mother so she can study their interactions in order to make her book more realistic. Meanwhile, the kidnapped editor begins to critique Evan's novel, which may be the book that the reader is reading. *The Seattle Times* calls *Tearjerker* a "sly little Möbius strip of self-reflective narrative invention."

Eugene Ionesco's *The Bald Soprano* (1950) has a Möbius-like twist at its conclusion. In the play, Mr. and Mrs. Smith invite Mr. and Mrs. Martin over for dinner. The play begins as a seemingly ordinary comedy on proper English manners. Mr. Smith is seated in his armchair and wears slippers. He smokes a pipe and reads a newspaper by the fireplace as he discusses food with Mrs. Smith.

But then weirdness ensues with irregular clock chimes and strange dialogue. In the beginning, the conversation makes sense, but the dialogue soon loses coherence and meaning, until the characters' responses seem to be random. The climax is like a dissonant symphony performed by musicians on LSD. The characters' inability to communicate leads to frustration and conflict. I don't think anyone reading the play could possibly understand what the last pages mean. Here is some sample dialogue toward the end of *The Bald Soprano*:

Mr. Martin: One doesn't polish spectacles with black wax.
Mrs. Smith: Yes, but with money one can buy anything.
Mr. Martin: I'd rather kill a rabbit than sing in the garden.
Mr. Smith: Cockatoos, cockatoos, cockatoos, cockatoos, cockatoos, cockatoos, cockatoos, cockatoos, cockatoos, cockatoos.
Mrs. Smith: Such caca, such caca, such caca, such caca, such caca, such caca, such caca, such caca, such caca.

The conclusion has a distinctly Möbius sort of loop to deepen the mystery: the characters reperform the play after exchanging roles. The final stage directions of the play read, "Mr. and Mrs. Martin are sitting like the Smiths at the beginning of the play. The play starts again with the Martins, who are saying exactly the same words as the Smiths in the first scene." The play has actually been performed with several variations on the twisted loop theme, so that the play oscillates with the same dialogue only with different couples saying the dialogue. Critics suggest that *The*

Bald Soprano shows how human conversation and other interactions have devolved into a collection of trite platitudes and how verbal mayhem erupts when proper English people lose their ability to communicate.

An easier to understand story, which is still filled with absurdity, is Danish writer Solvej Balle's *According to the Law* (1996). This book contains four interconnected stories that wrap around one another in a braided topological loop. The book starts with a Canadian biochemist who examines the brain of a young woman who has recently died of hypothermia and who has bequeathed her body to science. Next we meet Tanja, a Swiss law student who has paranormal powers that cause passersby to writhe in agony. We also encounter Danish mathematician Rene who wants to occupy as little volume as possible to become a human zero. Finally, Alette, a Canadian sculptor, dreams of merging with inanimate matter. She commits suicide and completes the Möbius strip by being the woman whose brain is being studied in the opening of the book.

In Stephen King's *Song of Susannah: Dark Tower VI*, King places himself in the book as a character. The gunslinger in the novel arrives in Maine in 1977 and hypnotizes a young horror writer, telling him he must finish the *Dark Tower* book series because the destiny of the world depends on it. King concludes the novel with a newspaper story about his death.

John Barth's *Lost in the Funhouse* has a foreword that explains how the book is "strung together on a few echoed and developed themes and [circles] back upon itself; not to close a simple] circuit like that of Joyce's *Finnegan's Wake*, emblematic of Viconian eternal return, but to make a circuit with a twist to it, like a Möbius strip, emblematic of—well, read the book."

The first Barth story in *Lost in the Funhouse*, called "Frame-tale," is literally a Möbius strip because it is a single page with the words "ONCE UPON A TIME THERE" written at one edge and "WAS A STORY THAT BEGAN" on the opposite side, with instructions for joining the ends to make a Möbius strip. Martin Gardner notes that the Doubleday edition of "Frame-tale," is designed to be read on an actual strip. The reader is told to cut the page along the dotted lines, then do a half twist to make a Möbius strip, on which one can endlessly read "Once upon a time there was a story that began once upon a time there was a story that began once upon a time there was a story that began . . ."

Barth himself said in a 1998 interview with Elizabeth Farnsworth on *NewsHour with Jim Lehrer,*

The tale is meant to be put on a Möbius strip, one of those guys that goes around—it's a circle with a twist, as is the book that follows it. . . . It's short on character, it's short on plot, but above all, it's short . . . and it does remind us of the infinite imbeddedness of the narrative impulse in human consciousness. I like to think if Scheherazade had had this little gadget, her problems would have been solved—the king would have gone to sleep, she could have started her novel, the end.

In a similar vein, Denise Duhamel's poem "Möbius Strip: Forgetfulness," in her 2005 book *Two and Two*, requires the reader to photocopy the poem and fashion it into a Möbius strip. The poem focuses on people with Alzheimer's disease and uses the strip to reinforce our impression of the distorted and fragmented nature of the afflicted person's mind.

Klein Bottle Literature Sampler

Novels and short stories have numerous references to Klein bottles. Paul J. Nahin's enigmatic story "Twisters," which appeared in the May 1988 edition of *Analog* magazine, begins with a Dr. Adams, a small-town physician, passing by a previously abandoned lot and noticing a doughnut shop that had not been there the day before. The kindly Dr. Adams reasons that with modern building techniques, it was at least *possible* that such a shop could be built in one day. Inside, he finds the usual assortment of doughnuts plus several "that had such curious twist" that at first he couldn't focus his eyes on an entire doughnut at once. He decides to buy a few of the twisted doughnuts. Later, while at his office, Dr. Adams finds that the doughnut absorbs all the coffee in his cup just by touching the liquid. And when he puts his ear near the doughnut, he hears a windy sound near its center. After much experimentation, Dr. Adams learns just how dangerous these "twisters" are as they absorb anything that takes a bite of them. "Apparently anything could pass through the gate . . . But it took the proximity of teeth (or more likely anything with calcium) to trigger the suction into overdrive." Adams determines that these doughnut twisters are Klein bottles and function as deadly traps made by the alien shopkeeper. Adams's goal for the remainder of the story is to make sure that no one takes a bite out of the tasty but deadly twister Klein bottles.

Martin Gardner's *Visitors from Oz* (1999) is a sequel to the Oz books in which Dorothy travels to New York City through a Klein bottle built from two Möbius strips by the same engineer who built the body of the Tin Man. While in New York, Dorothy appears on *The Oprah Winfrey Show*. Audience members, of course, think the Scarecrow and the Tin Man are just actors and not the real thing.

In Bruce Elliot's "The Last Magician" (1952), a magician uses a Klein bottle while performing for aliens. The trick turns out to be dangerous.

> Duneen was in real bad trouble. He was half in and half out of the Klein bottle. He was on the inside-outside, never-come-right side of the bottle. There he was, and there he is now. In the museum with all the other last things. And there he'll stay. They can't break the bottle because that would divide him. And since they can't break the bottle, there he will remain, not alive and not dead—suspended midway between here and there.

Andrew Crumey's *Möbius Dick* features a Möbius strip on the novel's cover. Crumey (pronounced "Croomey") has a Ph.D. in theoretical physics and is literary editor of *Scotland on Sunday*. In the novel, physicist John Ringer receives a text message on his "Q-phone" that simply says, "Call me: H." But who is H? Could "H" be his lover Helen from many years ago? This triggers his investigation into the development of new mobile phone technology taking place at a research facility in a Scottish village. During Ringer's adventures, the world transforms, and people experience amnesia, telepathy, false memories, and inexplicable coincidences. The plot is filled with psychoanalysis, inversions, cycles, and self-reflexive writing. Ringer wonders if coincidences are occurring with increasing frequency. If so, perhaps quantum experiments have caused the collapse of our universe's space-time continuum. Perhaps the twisted text of the novel comes from a parallel world.

When the reader discovers that a novelist named Harry Dick was writing a novel with a character named John Ringer, the reader begins to wonder which universe is real, or if "real" has any meaning at all. Throughout Möbius Dick, multiple stories coil around one another like trefoil knots. The funniest scene occurs when Ringer attends a woman's talk titled "Vicious Cycloids." During her presentation, the woman interprets a passage in *Moby Dick*, "with its facile relativism, its denial of objective certainty, its intellectual game playing"—a description that applies to *Möbius Dick* itself.

I hope you have enjoyed this brief introduction to movie and literature plots that feature nonorientable objects or that exhibit surprising, avant-garde loops. I look forward to hearing from you so that together we may catalogue additional examples of Möbius stories that both confound and delight. Let's conclude with three Mobius-like quotations that have always intrigued me:

"I am the thought you are now thinking."
—Douglas Hofstadter, *Metamagical Themas*

"As one goes through it, one sees that the gate one went through was the self that went through it."
—R. D. Laing, *The Politics of Experience*

He watched her for a long time
and she knew that he was watching her
and he knew that she knew he was watching her,
and he knew that she knew that he knew;
in a kind of regression of images
that you get when two mirrors face each other
and the images go on and on and on
in some kind of infinity."
—Robert Pirsig, *Lila*

◉ Ant Planet

Lisa is alone in her bedroom playing with bugs. She enjoys making mazelike struc-tures with leftover wires from her electrical experiment. If an object touches the wire, it rings a buzzer. Today, Lisa is experimenting with ants. The ants she places in these structures can only escape from certain locations without ringing the buzzer.

The ant prison mazes are of a peculiar type. Topologically speaking, they are Jordan curves, such as the one shown in figure 8.2, which is merely a circle that has been twisted out of shape. Recall that a circle divides any flat surface into two areas—inside and outside. Like a circle, Jordan curves have an inside and outside—and to get from one to the other, at least one line (wire) must be crossed.

Let's return to the ant story. Lisa is fantasizing about intelligent ants. One day, a

8.2
Ant trapped in a Jordan curve.

"prisoner" ant named Mr. Nadroj is able to accurately determine whether or not he is on the inside or the outside of the maze simply by poking his head over the wires and looking in one direction. What's the quickest way a creature can determine whether he is inside or outside the Jordan prison? How can you easily tell if the ant in the drawing can escape without actually trying to trace a path to the outside? (Turn to the solutions section for an answer.)

Life in Möbius Suburbia

🐜 *Swedesboro is one of those idiosyncratic, particular places that contradict the prevailing perception of South Jersey as little more than a Möbius strip of malls, an endless (if not relentless) agglomeration of big-box retail meccas with little in common but an area code and a propensity for traffic jams.*

—Kevin Riordan, *"South Jersey town debates identity,"*
The Courier-Post, *September 19, 2004*

🐜 *"You can be anywhere at all, any time. What do they call that thing?"*

"A Möbius strip," Henry said. "That's a nice idea. You could go back and visit your life anywhere and any time you wanted."

"It sounds like flypaper," Farlie said.

—Anne Rivers Siddons, Islands

CHAPTER 9

A FEW FINAL WORDS

What makes a great mathematician? A feel for form, a strong sense of what is important. *Möbius had both in abundance. He* knew *that* *topology was important. He* knew *that symmetry is a fundamental and* *powerful mathematical principle. The judgment of posterity is clear:* *Möbius was right.*

—Ian Stewart, *"Möbius's Modern Legacy," in* Möbius and His Band

FIG. 1

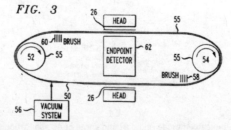

FIG. 2

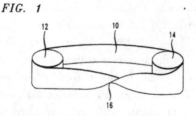

FIG. 3

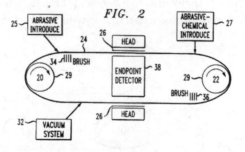

The Möbius Strip as a Launchpad

This completes our meandering survey of the Möbius strip in science, mathematics, and art. We've only touched the tip of the iceberg for most of these topics, but you should now have a better appreciation for the role of this looping, one-sided surface in a variety of disciplines. Sometimes I wonder why I am personally so compelled to contemplate the Möbius strip, and why so many people are delighted by its marvelous properties. Perhaps it is a metaphor for something eminently simple but surprising and difficult to predict. It is a ubiquitous symbol and, as in some novels, provides a vehicle to alter our minds and to see new worlds. It's the stuff of magic and the symbol of dreams.

I continue to be fascinated by the application of the Möbius strip in a range of technological inventions. Similar to the Reuleaux triangle— the triangle with curved sides that we discussed in chapter 4—these simple geometries did not find many practical applications until relatively late in humankind's intellectual development. Not until Franz Reuleaux (1829–1905) discussed his famous triangle (figure C.1), formed from the intersection of three circles at the corners of an equilateral triangle, did the curvy triangle begin to find numerous uses. Although Reuleaux wasn't the first to draw and consider such a curve, he was the first to demonstrate its constant-width properties and the first to use the triangle in numerous real-world mechanisms. The triangle's construction is so simple that modern researchers have wondered why no one before

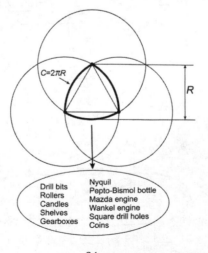

C.1

Reuleaux triangle (in bold) and some of its applications.

Reuleaux had exploited it use. The shape is a close relative of a circle because it has a constant width–the distance between two opposite points is always the same. Its circumference, $2\pi R$, is the same as for a circle, except that for the Reuleaux triangle, R is the length shown in figure C.1.

As I drift off to sleep at night, I imagine glittering Möbius strips and Reuleaux triangles while I contemplate new inventions and think about the shape of our universe. Our nature is to dream, to search, and to wonder about our place in a seemingly lonely cosmos. Perhaps this is a reason that philosophers and writers have speculated about universes and higher dimensions shaped like Möbius strips, and what their inhabitants might be like. For many young prospective scientists, the Möbius strip is a launchpad to more sophisticated geometries and topological exploration.

Many cosmological models have been devised in which our universe curves through 4-space in a way that could, in theory, be tested. For example, Einstein suggested a universe model in which a spaceship could set out in any direction and return to its starting point. In this model, our 3-D universe is treated as the hypersurface of a huge hypersphere. Going around it would be comparable to an ant walking around the surface of a sphere. In other universe models, our universe is a hypersurface that twists through 4-space like a Klein bottle or a 3-torus, a doughnut wrapped in three dimensions.

Using various satellites, astronomers now actively search for evidence of the universe's shape by studying temperature fluctuations in deep space. Although recent evidence suggests that the nearby regions of our universe may be quite ordinary, no one knows how the entire cosmos may be shaped.

Mathematicians dating back to Georg Bernhard Riemann (1826–1866) have studied the properties of multiply connected spaces in which different regions of space and time are spliced together. Physicists who once considered this an intellectual exercise for armchair speculation are now seriously studying advanced branches of mathematics to create practical models of our universe and to better understand the possibilities of parallel worlds, travel using wormholes, and methods for manipulating time. Even if these odd universes are unlikely or difficult to ascertain given current technology, the shapes in this book make physicists alert to numerous possibilities to consider for topological triage– a process in which models are sorted and considered in terms of their likeness to elucidate the nature of the universe. Both lay people and scientists have become more aware about what it means to visualize an abstract object or twisted space.

Zen Buddhists have developed questions and statements called koans that function as a meditative discipline. Koans ready the mind so that it can entertain new intuitions, perceptions, and ideas. Koans cannot be answered in ordinary ways because they are paradoxical; they function as tools for enlightenment because they jar the mind. Similarly, the contemplation of the Möbius strip is replete with koans, and that is why this book teaches you with so many different topics, although space limitations did not permit us to explore any single topic in great depth. The Möbius strip is a koan for scientific minds.

Eternal Sunshine

For me, some of the most interesting areas for Möbioid behavior and Möbius koans occur in literature, where the Möbius strip is a metaphor for looping plots. We've discussed several examples in chapter 8. The Möbioid movies and stories function partly as metaphysical explorations. Often they are dark, self-enclosed, with a special dreamlike logic to help us transcend individual consciousness.

We can conclude with another Möbius-structured plot. In the haunting movie *Eternal Sunshine of the Spotless Mind,* Joel and Clementine decide to erase their memories of each other after they fall out of love. Inevitably, they meet again, retaining vestiges of their memories, and fall in love again.

Much of the movie takes place within Joel's mind. When Joel is in the middle of the memory-erasing procedure, he becomes aware that his memories of the woman he loved are disappearing from his mind, and he wants the procedure stopped. His herculean task is to devise ways to protect as much of his memory of Clementine as he can, and to find a way to escape from the procedure despite being in a dreamlike state.

Near the end of the movie, the viewer returns to the movie's start, where one sees the opening scene in light of new knowledge of the characters' predicaments. Möbius ribbons are everywhere in the story. In some of the dreamlike scenes during the mind-erasure procedures, Joel chases Clementine down a street only to find the street looping back on itself, and he frustratingly keeps finding Clementine running behind him. Thus, not only do we have a Möbius plot in which the end of the movie returns to the beginning with the characters changed by vestigial memories, we also have the insertion of looping realms within the recesses of their minds as Joel and Clementine fight to remember the love they once shared, to understand what is real and what is fantasy. Does an eternal bond continually unite and reunite Joel and Clementine as they loop back and find that what they're seeing all happened earlier and is just now about to be erased from their minds?

Perhaps at the end of the movie, Joel and Clementine have found a way to leave the surface of the Möbius strip. Instead of running from each other, love brings together their imperfect personalities, making the line between reality and fantasy unimportant. They cherish their time together and live every moment in the "now," relishing their dreams, living close to each other, knowing that their dreams, at any time, could be cleansed from their minds forever.

Simple Math

Many of the Möbius puzzles in this book are of interest to recreational mathematicians and mathematical amateurs, groups of enthusiasts whose members have had strong records in making important mathematical discoveries. In 1998, self-taught inventor Harlan Brothers and meteorologist John Knox developed an improved way of calculating a fundamental constant e (often rounded to 2.718). Studies of exponential growth—from bacterial colonies to interest rates—rely on e, which can't be expressed as a fraction and can only be approximated using computers. Knox demonstrated that amateurs continue to make strides in mathematics and can help find more accurate ways of calculating fundamental mathematical constants. Perhaps you will one day discover some remarkable new property of the Möbius strip or invent a new toy based on its peculiar properties.

Another "beginner" who made substantial contributions to mathematics was Marjorie Rice, a San Diego housewife and mother of five, who was working at her kitchen table in the 1970s when she discovered numerous new geometrical patterns that professors had thought were impossible. Rice had no training beyond high school, but by 1976 she had discovered fifty-eight special kinds of pentagonal tiles, most of which were previously unknown. Her most advanced diploma was a 1939 high school degree for which she had taken only one general math course. The moral of the story? It's never too late to enter fields and make new discoveries. Another moral: Never underestimate your mother!

The idea that very simple yet profound mathematics can still be discovered today is not as far-fetched as it sounds. For example, mathematician Stanislaw Ulam, in the mid to late twentieth century, was bursting with simple but novel ideas that quickly led to new branches of mathematics such as those that focus on cellular automata theory and the Monte Carlo method. As Martin Gardner points out in "The Adventures of Stanislaw Ulam" (1976), "Over and over again, Ulam has obtained profound results in fields about which he knew little. Perhaps because of that, he was able to see the problems in fresh ways."

Another example of simplicity and profundity is Penrose tiling–the pattern of tiles we discussed in chapter 7 that is made with just two shapes of tile and was discovered as recently as 1974 by Roger Penrose. These tiles can completely cover an infinite surface in a pattern that is always nonrepeating (aperiodic). Aperiodic tiling was first considered merely a mathematical curiosity, but physical materials were later found in which the atoms were arranged in the same pattern as a Penrose tiling, and now the field has an important role in chemistry and physics. We should also consider the intricate and strikingly beautiful behavior of the Mandelbrot set, a complicated fractal object described by a simple formula, $z = z^2 + c$, and unearthed at the end of the twentieth century (figure C.2).

C.2

An example of a fractal, a shape with infinite detail that is spawned from a simple mathematical formula.

Computers will probably facilitate future discoveries of startling characteristics of seemingly simple mathematics. Returning to the striking
example of the Mandelbrot set, Arthur C. Clarke in *The Ghost from the
Grand Banks* once noted,

> In principle [the Mandelbrot set] could have been discovered as
> soon as men learned to count. But even if they never grew tired,
> and never made a mistake, all the human beings who have ever
> existed would not have sufficed to do the elementary arith-
> metic required to produce a Mandelbrot set of quite modest
> magnification.

Dr. Mandelbrot himself discussed his discovery of the set in a 2004 *New
Scientist* interview:

> Its astounding complication was completely out of proportion
> with what I was expecting. Here is the curious thing: the night I
> saw the set, it was just wild. The second night, I became used to
> it. After a few nights, I became familiar with it. It was as if
> somehow I had seen it before. Of course, I hadn't. No one had
> seen it. No one had described it. The fact that a certain aspect of
> its mathematical nature remains mysterious, despite hundreds of
> brilliant people working on it, is the icing on the cake to me.

Unlike the Mandelbrot set, the Möbius strip didn't require a computer to help reveal its profundity. Thus, the Möbius strip is the ultimate
metaphor for something simple, yet profound—something anyone could
have discussed centuries prior to its discovery, but didn't. The Möbius
strip is a metaphor for magic and mystery, and a perpetual icon that stim-
ulates us to dream new dreams and look for depths even in seemingly
shallow waters.

✒ Ambiguous Ring

*Figure C.3 shows an ambiguous ring. Why or why not is this the same as a Möbius
strip? (Turn to the solutions section for an answer.)*

BRIAN C. MANSFIELD

C.3
Ambigious ring. Is this a Möbius Strip?

Möbius in Business and Government

🎤 *Government today is organized as kind of a Möbius strip, with mistrust flowing round and round.*
—Philip K. Howard, The Collapse of the Common Good:
How America's Lawsuit Culture Undermines Our Freedom

🎤 *Personal names also create a unique, self-reinforcing benefit loop—a kind of marketing Möbius strip. When people whose names are included in the firm name get good publicity, the entire firm's reputation is enhanced.*
—Harry Beckwith, What Clients Love: A Field Guide to Growing Your Business

🎤 *I watched agog as sniggering reports of CIA torture were passed off as something normal on TV amid Möbius-strip assurances that . . . the President is blameless . . .*
—Michael Gilson-De Lemos, "MG's Most Controversial Article Yet," Citizens for
Legitimate Government

🎤 *Those consequences in a Möbius strip world where everything folds back into our own lives are not just "out there" but "in here," in our souls, where the corrosive acid of self-deceit challenges the American belief that we are good or better or different.*

—Richard Thieme, *"I Was a Victim of the KGB,"* Common Dreams News Center

FIG. 2A

FIG. 2B

SOLUTIONS

Chapter 1

The treadmill seems to be locked. Assuming the good doctor is trying to run forward, each axle in the machine should turn counterclockwise as seen from the reader's point of view. However, the figure eight belt, in order to work properly, needs to turn each of its axles in a different direction.

If the figure eight belt is replaced by a Möbius strip belt in the form of a loop (not a figure eight), the machine should work and turn the axles to which it is attached in the same direction. In fact, such belts are generally superior to ordinary belts because they can wear half as quickly because the belt presents both "sides" of the rubber to the axles.

Chapter 2

To solve this knotty problem, consider that there are two possible crossings at each intersection point. This means that there are $2 \times 2 \times 2 = 8$ possible sets of crossings. Of all these possibilities, only two create a knot. (Test this for yourself using a loop of string.) Thus, the probability of having a knot is one in four. Don't bet on it happening!

Figure A.1 shows another possible rope configuration. What are the odds that it forms a knot? Does the probability of knot formation increase with increasing numbers of intersection points? What does this say about Murphy's Law—that ropes and strings and electrical cords always seem to get tangled when thrown in a jumble in your garage?

A.1
Another possible rope configuration. What are the odds that it forms a knot?

Chapter 3

Figure A.2 is one solution to "The Möbius Maze" by Dave Phillips.

A.2
One solution to "The Möbius Maze" by Dave Phillips.

Chapter 4

Figure A.3 shows one solution. Many of my genius friends have told me that this puzzle was impossible to solve. However, if my friends looked at the puzzle a day later, they usually could solve it on the second day.

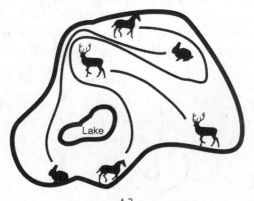

A.3
One solution to the Noah's Ark puzzle.

Chapter 5

For the squiggle puzzle, note that if you draw a map on a plane using a continuous line, not taking your pencil off the paper and returning to the starting point, you only need two colors to produce a map in which any regions with a common boundary line have different colors. Figure A.4 is one example of such a coloring. Try it with other patterns!

A.4
A solution to the squiggle map puzzle.

For the pyramid puzzle, it turns out that the second view is incorrect. First, let's look at views two and four. Notice that the missing color is green for both. This means green must be at the hidden base in these views. Thus, two and four can't both be right. One of them must be wrong, and therefore the first, third, and fifth views must be correct.

Now, let's examine the third view. Try to imagine the green facet as the base for view three. That means the sides of the pyramid are red, purple, and yellow. This is the same as the pyramid in view four. This leaves us with view two being the incorrect view.

Chapter 6

Figure A.5 is one way you can transform the linked rings to the unlinked rings.

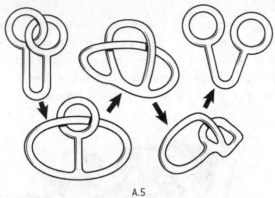

A.5

One way in which you can transform the linked rings to the unlinked rings without cutting a ring. (After David Wells, *The Penguin Dictionary of Curious and Interesting Geometry*.)

Let's conclude with yet another loop problem, and I'll give you the answer so as not to torture your brain further. In figure A.6, it's possible to transform the three interlocked loops at the left of the figure so that one loop is unchained from the other.

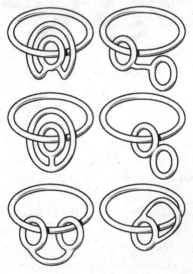

A.6

Another wonderful transformation without breaking loops. The three loops at the left transform to the configuration at the bottom right so that one loop is unchained from the other. (After David Wells, *The Penguin Dictionary of Curious and Interesting Geometry*.)

Chapter 7

Figure A.7 indicates the path to travel to solve the torus maze. Simply travel from *1* to *2* to *3* to *4*. Figure A.8 shows the solution to the Klein bottle maze. Travel from *1* to *2*, and you're finished!

A.7
Solution for the torus maze.

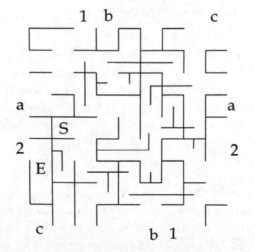

A.8 Solution for the Klein bottle maze.

Regarding the devil configurations, the constructions are equivalent to a twist with the addition of a single connecting strip, as shown in figure A.9. If one or both twist sequences contain an odd number of half twists, then only one piece results when the figure is cut along the dotted line. If each sequence of twists contains an even number of half twists, two separate and interwined pieces result.

A.9
A twisted strip with the addition of a single connecting strip.

The solution to the problem illustrated in figure 7.44 is the large square band framing figure 7.44. The solution is the same no matter how many half twists you include in one arm. A flat square band of paper always results! I learned about these puzzles in James Tanton's *Solve This: Math Activities for Students and Clubs* and Martin Gardner's *Mathematical Magic Show.*

Chapter 8

The quickest way Mr. Nadroj can tell whether he is inside or outside the Jordan curves is to count the number of times an imaginary line drawn from his body to the outside world crosses a wire. Figure A.10 shows several sample lines. If the straight line crosses the curve an *even* number of times, the ant is outside the maze; if it crosses an *odd* number of times, the ant is inside.

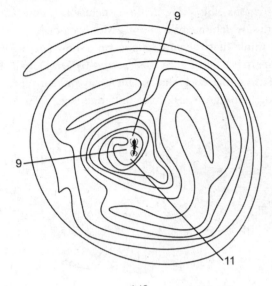

A.10
How to easily determine if the ant is within the prison without tracing paths.

Back in the real world, French mathematician Marie Ennemond Camille Jordan (1838–1922) offered a proof of the same rules for determining the inside and outside of these kinds of curves. (The proof was corrected in 1905 by Oswald Veblen.) Jordan was originally trained as an engineer.

Note that a Jordan curve is a plane curve that is a deformed circle, and it must be simple (the curve cannot cross itself) and closed (it must have no endpoints and must also completely enclose an area). On a plane or sphere, Jordan curves have an inside and outside—and to get from one side to the other, at least one line must be crossed. However, on a torus, Jordan curves do necessarily require a line crossing.

Conclusion

I believe this object can be classified as an optical illusion or an "impossible object" in the spirit of other famous impossible objects with names such as the Freemish crate, the Penrose staircase (often drawn by M. C. Escher and on which you can walk "upstairs" forever), the Penrose tribar (which has three cylindrical prongs arranged in a strange way), the Penrose triangle, and the ambihelical hexnut. You can find many more similar objects on the Web.

The ambiguous ring show in the conclusion (figure C.3) seems to have two sides, which means it is not a Möbius loop.

Do you think that the ambiguous ring shown in figure A.11 has Möbius properties? This figure was created by Dr. Donald E. Simanek, professor of physics at Lock Haven University of Pennsylvania. Does the Penrose triangle in figure A.12 have only one face?

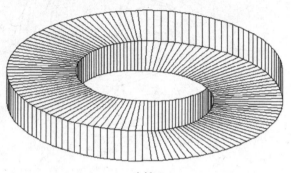

A.11
Ambiguous ring by Dr. Donald E. Simanek.

A.12
Penrose triangle.

If you use your hand to cover the left or right third of figure A.11, the figure seems "conflict-free." When you look at the left side of the ring, everything seems normal, consistent with a washer seen from above. When you look at the right of the ring, everything seems consistent with a washer seen from below. But when you observe all of the ring at once, your analytical brain kicks in and says "impossible!"

Your journey ends with a maze by Dave Phillips.
Help one robot find the other by traveling the shortest route.
The ends of spirals are dead ends.
No climbing over the edges, of course.

Möbius Strip in Language

🔲 But "not sure" is exactly how a lot of Vietnam soldiers, Kerry included, felt about the war mission itself. Kerry wrote this passage in such a way that you can take it the other way, too, if you feel like it. This is Möbius-strip rhetoric, which reads, "I'm not sure I'm coming home" and "I'm not sure I'm doing the right thing" on the same side of a one-sided strip, and as plain an example as you will ever see of a politician talking out of both sides of his mouth.

—Matt Taibbi, "Mere Words," FreezerBox.com

🔲 At two hours without interruption, though, the Möbius-strip dialogue can grow disorienting and it's possible to miss the unobtrusive conclusion altogether.

—"Shimmer Traverse Theatre," Edinburgh Financial Times

REFERENCES AND APPENDIX

The second-quantized fermionic vacuum state of the $G = SU(2)$ and $r = 2_L$ chiral Yang-Mills theory in the Hamiltonian formulation (temporal gauge $W_0 = 0$) then has a *Möbius bundle* structure over a specific non-contractible loop of x^3-independent static gauge transformations.

–F. R. Klinkhamer, "Z-string Global Gauge Anomaly and Lorentz Non-Invariance," *Nuclear Physics B*, 1998

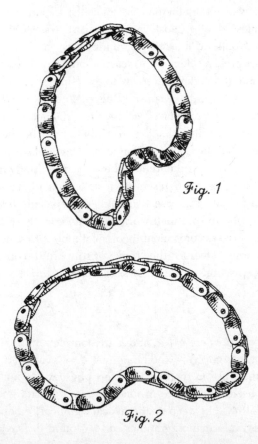

Fig. 1

Fig. 2

I've compiled the following reference list that identifies much of the material I used to research and write this book. It includes information culled from books, journals, and Web sites. As many readers are aware, Web sites come and go. Sometimes they change addresses or completely disappear. The Web site addresses listed here provided valuable background information when this book was written. You can, of course, find numerous other Web sites relating to the Möbius strip by using search tools such as the ones provided at www.google.com.

If I have overlooked an interesting mathematical puzzle, person, reference, or factoid relating to Möbius that you feel has never been fully appreciated, please let me know about it. Just visit my Web site, www.pick over.com, and send me an e-mail explaining the idea and how you feel it influenced the world. In the interest of space, I have intentionally not covered more advanced mathematical concepts, including Möbius nets, Möbius dualities, Möbius transforms, Möbius statics, Möbius transformations, Möbius groups, Möbius inversion formulas, and Möbius bundles. If readers have a pressing demand to learn about these subjects, perhaps I will write a future book devoted solely to these intricate topics.

In the meantime, you may consult Roger Penrose's *The Road to Reality: A Complete Guide to the Laws of the Universe* (2005) for related Möbius delights, including an introduction to Möbius fiber bundles. Generally speaking, a fiber bundle is a space that locally resembles a product of two spaces but may possess a different global structure. Mathematical drawings of fiber bundles often resemble a collection of hairs (the fibers) growing from a scalp (the base manifold)—as depicted at MathWorlds's bundle Web site: http://mathworld.wolfram.com/FiberBundle.html. Fiber bundles serve as convenient theoretical tools for particle physicists.

To give readers a feel for the "look" of some of the other advanced Möbius concepts, consider that a *Möbius transformation* is a function of the form

$$f(z) = \frac{az+b}{cz+d}$$

where $ad \pm bc$ and where a, b, c, and d are complex numbers. The point $z = -d/c$ is mapped to $f(z) = \infty$. The point $z = \infty$ is mapped to $f(z) = a/c$. Aside from their use in mathematics and physics, Möbius transformations can be used by artists to produce stunning fractal images (figures R.1, R.2, R.3, R.4). The deep mathematical significance behind many of these Möbius-tranformation graphics can be found in David Mumford,

Caroline Series, and David Wright's *Indra's Pearls: The Vision of Felix Klein* (2002). The shapes are fractals produced by iteration (repetition) of Möbius transformations and their inverses. The details in the figures continue for many magnifications—like endlessly nested Russian dolls.

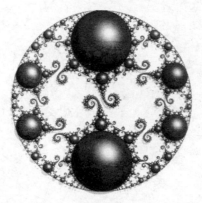

R.1

A Kleinian group image—a limit set generated by Möbius transformations of the form $z \rightarrow (az + b)/(cz + d)$. More particularly, this fractal image was generated with two Möbius transformations and their inverse transformations. This iterative process will repeatedly displace an initial point in the complex plane. The resultant set of points forms the limit set, represented graphically in this figure. No matter how often and in what order the displacements are repeated, the new points fall somewhere on the figure's curved shapes. Möbius transformations will transform circles to circles, and this property yields the sphere-like objects in the image. (Artwork by Jos Leys, www.josleys.com.)

R.2	R.3
Same as R.1, with different values of $a, b, c,$ and d. (Artwork by Jos Leys, www.josleys.com.)	A graphical experiment using the Möbius transformation. (Artwork by Ed Pegg, Jr., www.mathpuzzle.com)

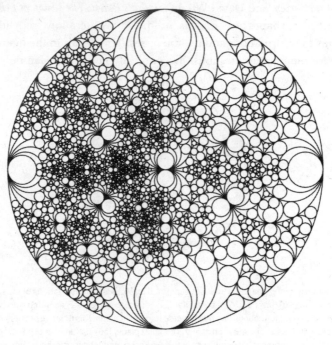

R.4

Asmus Schmidt's complex continued fraction algorithm uses Möbius transforms to generate ever-finer tessellations of the plane. Doug Hensley's picture of part of the fifth tessellation illustrates the striking reservoir of shapes and patterns that are woven into the fabric of mathematics.

The *Möbius transform Tf* of a function *f* defined on the positive integers is represented as

$$(Tf)(n)=\sum_{d\mid n}f(d)\mu(n/d)=\sum_{d\mid n}f(n/d)\mu(d)$$

where μ is the usual Möbius function, and the notation $d \mid n$ indicates that d is a divisor of n. The function *Tf is* also called the Möbius inverse of *f.*

Using the *Möbius inversion formula,* if $g(n)$ and $f(n)$ are arithmetic functions satisfying

$$g(n)=\sum_{d\mid n}f(d) \text{ for every integer} \geq 1$$

then

$$f(n)=\sum_{d\mid n}g(d)\ \mu(n/d) \text{ for every integer} \geq 1$$

where μ is the usual Möbius function, and the notation $d \mid n$ indicates that d is a divisor of n.

In *Möbius's duality* for three-dimensional space, each point corresponds to a plane, and vice versa. Jeremy Gray in *Möbius and His Band* notes that "the anti-symmetric case is new, and is one of Möbius's finest discoveries. The discovery that in odd-dimensional spaces there is a new kind of duality, not associated with quadrics, is due to Möbius and arose from his study of geometrical mechanics."

Readers may consult www.wikipedia.org or http://mathworld.wolfram.com for different notations for these formulas. These Web sites are regularly updated as readers discover additional properties and applications for these Möbius concepts. Fauvel, Flood, and Wilson's *Möbius and His Band* provides further insight.

Deliciously complicated quotations can be pulled from the Web when searching for references to the Möbius bundle. I leave you with this gem to accompany the quotation that begins this section:

The Thom complex of the Möbius band is the projective plane, and MD is its suspension spectrum, i.e. SZ_2. The transformation $\theta_D\colon MD \to HZ_2$ can now be identified with the inclusion of the (stable) 1-skeleton, i.e., the mod 2 Hurewicz map. . . . Denote by ρ/I the *Möbius bundle* which is the 1-dimensional vector bundle constructed by gluing the trivial bundles over $[0, 1/2]$ and $[1/2, 1]$ by multiplication by -1 in the fiber over $1/2$. Thus, although ρ/I is the trivial bundle, it is equipped with opposite trivializations on the two halves of I. Denote by $\hat{\rho}/I$ the stable version. We call $\hat{\rho}/I$ the (stable) Möbius twist.

—Roger Fenn, Colin Rourke, Brian Sanderson,
"James Bundles," 2003

Readers continue to send me creative mazes made on Möbius worlds floating in space. Figure R.5 is a gem from master maze maker, Dave Phillips. He wrote this to me: "Find the path that the four flies take if they all travel the same route without meeting, and without retracing their path until they reach their original position. Keep track of which side of the path you are on."

R.5

Möbius fly maze by Dave Phillips. Find the path that the four flies take if they all travel the same route without meeting and without retracing their path until they reach their original position. Keep track of which side of the path you are on.

Möbius Strip in Aesthetics

🪰 *What had seemed like a linear progression was really a kind of Möbius strip: The progression of art began at Lascaux only to end, some 15,000 years later, with artists aspiring to paint like cavemen. Now, after the end of art, anything goes.*

—Natasha Degen, *"The Philosophy of Art: A Conversation With Arthur C. Danto,"* The Nation

🪰 *Served in a small, double-handled cast-iron pot, the ingredients were arranged in harmonious Japanese symmetry. Ivory colored slices of creamy textured tofu fanned half the circumference of the pot, met by Möbius-strip-like curls of gelatinous fish cake, followed by round daikon slices that looked like pale apple rings, and ending with spears of baby corn.*

—Lorraine Gengo, *"My Traves With Sukiyaki,"* Fairfield County Weekly

READINGS BY CHAPTER

Introduction

Fauvel, John, Raymond Flood, and Robin Wilson. *Möbius and His Band: Mathematics and Astronomy in Nineteen-Century Germany*. New York: Oxford University Press, 1993.

Gardner, Martin. *Hexaflexagons and Other Mathematical Diversions: The First Scientific American Book of Mathematical Puzzles and Games*. 1959. Reprint, Chicago: University of Chicago Press, 1988.

Smith, Bruce. "Brewer marketing new energy beer." *Columbia (SC) State*, November 21, 2004. http://www.thestate.com/mld/thestate/news/local/10240117.htm. (On Möbius beer.)

Chapter 1

Gardner, Martin. *Mathematics, Magic and Mystery*. New York: Dover, 1956).

Reamer, Eric. "Ministry Trick of the Month: Möbius Madness." Eric Reamer's Illustrated Illusions. http://www.illustratedillusions.com/trick26.htm.

Regling, Dennis. "Make It Yourself Magic Loops." Bella Online. http://www.bellaonline.com/articles/art5655.asp.

Chapter 2

Adams, Colin. "Why knot: knots, molecules and stick numbers."Plus. http://plus.maths.org/issue15/features/knots/.

Bows, Alice. "Non-collapsing knots could reveal secrets of the Universe." Institute of Physics.

http://www.innovations-report.com/html/reports/physics_astronomy/
report-8726.html. (See also http://physics.iop.org/IOP/Press/PR3002.html.)

Haken, Wolfgang. "Theorie der Normalflachen." *Acta Mathematica* 105
(1961): 245–375.

Gardner, Martin. *Mathematical Magic Show: More Puzzles, Games, Diversions,
Illusions and Other Mathematical Sleight-of-Mind from Scientific American.*
New York: Vintage, 1978.

Kornbluth, Cyril. "The Unfortunate Topologist" (limerick). *Magazine of Fantasy and Science Fiction.* Reprinted in *Fantasia Mathematica.* Edited by Clifton
Fadiman. New York: Simon and Schuster, 1958. (In 1958, Cyril M. Kornbluth died of a heart attack at age thirty-five. He was famous for his science
fiction collaborations with Frederik Pohl on *The Space Merchants* (1953),
Search the Sky (1954), and *Wolfbane* (posthumously published in 1959).

Leys, Jos. "Computer generation of Celtic Knots." Fractals by Jos Leys.
http://www.josleys.com/creatures36.htm.

Leys, Jos. "Klein Bottles, Trefoil Knots, and Beyond." Fractals by Jos
Leys. http://www.josleys.com/creatures48.htm.

Menasco, William W. "A Circular History of Knot Theory." William
Menasco's Home Page. http://www.math.buffalo.edu/~menasco/

Nyhart, Lynn K. "Economic and Civic Zoology in Late Nineteenth-Century Germany: The 'Living Communities' of Karl Möbius." *Isis* 89
(1998): 605–630. (Alas, when I asked Dr. Nyhart if Karl Möbius could
possibly be related to the mathematician Möbius, her answer was no.)

Perko, Kenneth A., Jr. "On the classification of knots." *Proceedings of the
American Mathematical Society* 45, no. 2 (August 1974): 262–266.

Peterson, Ivars. *Islands of Truth: A Mathematical Mystery Cruise.* New
York: Freeman, 1990.

Robinson, John. "Symbolic Sculpture."
http://www.bangor.ac.uk/cpm/SculMath/main.htm.

Scharein, Robert G. "Perko Pair Knots." Center for Experimental and Constructive Mathematics. http://www.cecm.sfu.ca/~scharein/projects/perko/

Sossinsky, Alexei. *Knots: Mathematics with a Twist.* Cambridge: Harvard University Press, 2002.

Taylor, William. "A deeply knotted protein structure and how it might fold," *Nature* 406 (2000): 916–919. (See also Eric Martz, "Knots in Proteins," http://www.umass.edu/microbio/chime/knots/.
See a preprint of the paper on *Nature*'s Web site, http://www.nature.com/cgi-taf/DynaPage.taf?file=/nature/journal/v406/n6798/abs/406916a0_fs.html.)

"U. S. Navy *Bluejacket's Manual*," http://usselectra.org/bjm/bjmie.html. The knot images come from the 1943 edition published by the U.S. Navy Institute, Annapolis, Maryland. A more recent edition, which is more easily acquired, is Thomas J. Cutler, *The Bluejacket's Manual* (Centennial Edition) (Annapolis, Maryand: Naval Institute Press, 2002)

Chapter 3
Note: Although János Bolyai and Nikolai Lobachevsky seemed to have developed hyperbolic geometry independently and at the same time, legends suggest that both learned about it indirectly from Carl Friedrich Gauss, who worked in this area. Although Gauss never published the material during his lifetime, he told the father of Bolyai and a colleague of Lobachevsky about his findings. Most mathematicians with whom I consulted say that this legend is based on just two letters that Gauss wrote to Bessel, but the information in the letters is not explicit.

Abbott, David, ed. *Mathematicians* (*The Biographical Dictionary of Scientists*). New York: Peter Bedrick Books, 1985.

Crowe, Michael. "August Ferdinand Möbius." In *Dictionary of Scientific Biography*. Edited by Charles Gillispie. New York: Charles Scribner, 1974.

Fauvel, John, Raymond Flood, and Robin Wilson. *Möbius and His Band: Mathematics and Astronomy in Nineteenth-Century Germany.* New York: Oxford University Press, 1993.

Fritsch, Rudolph. "Möbius Biography." http://www.mathematik.uni-muenchen.de/~fritsch/Moebius.pdf.

Katz, Eugenii. "The Charles Martin Hall and Aluminum." http://www.geocities.com/bioelectrochemistry/hall.htm.

Möbius, August. *Gesammelte Werke.* Edited by Richard Baltzer, Felix Klein, and Wilhelm Scheibner. 4 vols. Leipzig: reprinted, Dr. Martin S"ndig oHg, Wisbaden, 1967.

O'Connor, John J., and Edmund F. Robertson. August Ferdinand Möbius. http://www-history.mcs.st-andrews.ac.uk/Mathematicians/Möbius.html.

Pickover, Clifford. *Calculus and Pizza: A Math Cookbook for the Hungry Mind.* New York: Wiley, 2003.

Pickover, Clifford. *A Passion for Mathematics: Numbers, Puzzles, Madness, Religion, and the Quest for Reality.* New York: Wiley, 2005.

Querner, Hans. "Karl Augus Möbius." In *Dictionary of Scientific Biography.* Edited by Charles Gillispie. New York: Charles Scribner, 1974.

Weisstein, Eric. "Möbius, August Ferdinand." Eric Weisstein's World of Scientific Biography. http://scienceworld.wolfram.com/biography/Moebius.html.
(This reference points out that a fascinating numerical result of Möbius's barycentric calculus is that the probability that five points randomly chosen in the projective plane lie on a hyperbola is infinitely greater than the probability that they lie on an ellipse.)

Yaglom, Isaak Moiseevich. *Felix Klein and Sophus Lie: Evolution of the Idea of Symmetry in the Nineteenth-Century. P.39.* Boston: Birkhauser, 1988.

Chapter 4

Ajami, D., O. Oeckler, A. Simon, and R. Herges. "Synthesis of a Möbius aromatic hydrocarbon." *Nature* 426 (December 18, 2003) 819–821.

Albrecht-Gary, A. M., C. O. Dietrich-Buchecker, J. Guilhem, M. Meyer, C. Pascard, J. P. Sauvage. "Dicopper (I) Trefoil Knots: Demetallation

Kinetic Studies and Molecular Structures." *Recueil des Travaux Chimiques des Pays-Bas* 112 no. 6 (1993): 427–428.

Beavon, Rod. "Chirality." http://www.rod.beavon.clara.net/chiralit.htm

Billingsley, Patrick. "Prime numbers and Brownian motion." *American Mathematical Monthly* 80, no. 1099 (1973). http://www.maths.ex.ac.uk/~mwatkins/zeta/wolfgas.htm (Addresses the use of the Möbius function in quantum field theory.)

"The Cost of Ideas." *Economist* 313, no. 8401 (November 13, 2004): 71. http://economist.com/opinion/displayStory.cfm?Story_id=3388936. (On the state of patents in the twenty-first century.)

Dietrich-Buchecker, C. O., J. Guilham, C. Pascard, J.-P. Sauvage, "Structure of a Synthetic Trefoil Knot Coordinated to Two Copper(I) Centers," *Angewandte Chemie* (International Edition English) 29, no. 1154 (1990).

Dietrich-Buchecker, C. O., and J.-P. Sauvage. "A Synthetic Molecular Trefoil Knot." *Angewandte Chemie* (International Edition English) 28 (1989): 189–192.

Dietrich-Buchecker, C. O., J.-P. Sauvage, J.-P. Kintzinger, P. Maltese, C. Pascard, J. Guilhem. "A Di-copper(I) Trefoil Knot and Its Parent Ring Compounds: Synthesis, Solution Studies and X-ray Structures." *New Journal of Chemistry* 16 (1992): 931–942.

Dietrich-Buchecker, C. O., J.-F. Nierengarten, J.-P. Sauvage, N. Armaroli, V. Balzani, L. De Cola. "Dicopper(I) Trefoil Knots and Related Unknotted Molecular Systems Influence of Ring Size and Structural Factors on their Synthesis and Electrochemical and Excited-state Properties." *Journal of the American Chemical Society* 115 (1993): 11237–11244.

Dietrich-Buchecker, C. O, J.-P. Sauvage, A. De Cian, J. Fischer. "High-yield Synthesis of a Dicopper(I) Trefoil Knot Containing 1,3-phenylene Groups as Bridges Between the Chelate Units." *Chemical Society, Chemistry Communications* 19 (1994): 2231–2232.

Du, S. M. and N. C. Seeman. "The Construction of a Trefoil Knot from a DNA Branched Junction Motif." *Biopolymers* 34 (1994): 31–37.

Hoffman, Paul. *Archimedes' Revenge: The Joys and Perils of Mathematics.* New York: Norton, 1988.

Joy, Linda. "Knot to be Undone, Researchers Discover Unusual Protein Structure." National Institute of Health. http://www.nih.gov/news/pr/nov2002/nigms-26.htm.

Martín-Santamaría, Sonsoles, and Henry S. Rzepa. "Twist Localisation in Single, Double and Triple Twisted Möbius Cyclacenes." *Journal of the Chemical Society.* 2, no. 12 (2000): 2378–2381. http://www.rsc.org/suppdata/P2/B0/B005560N/b005560n.htm.

Möbius, August. *Gesammelte Werke.* Edited by Richard Baltzer, Felix Klein, and Wilhelm Scheibner. 4 vols. Leipzig: reprinted, Dr. Martin Sändig oHg, Wisbaden, 1967.

Rzepa, Henry S. "Molecular Möbius Strips and Trefoil Knots." http://www.ch.ic.ac.uk/motm/trefoil/

Sauvage, Jean-Pierre. "Interlocking Rings and Knots at the Molecular Level." *Leonardo* 30, no. 4, (August 1997): 276–277. http://mitpress.mit.edu/catalog/item/default.asp?tid=5005&ttype=6

Spector, Donald. "Supersymmetry and the Möbius Inversion Function." *Communications in Mathematical Physics* 127 (1990): 239.

Tanda, Satoshi, Tsuneta Taku, Okajima Yoshitoshi, Inagaki Katsuhiko, Yamaya Kazuhiko, and Hatakenaka Noriyuki. "A Möbius Strip of Single Crystals." *Nature* 417, no. 6887 (May 23, 2002): 397–398.

Walba, David, Rodney Richards, and R. Curtis Haltiwanger. "Total Synthesis of the First Molecular Möbius Strip." *Journal of the American Chemical Society* 104 (1982): 3219–3221.

Walba, D. M., T. C. Homan, R. M. Richards, and R. C. Haltiwanger.

"Topological Stereochemistry. 9. Synthesis and Cutting in Half of a Molecular Möbius Strip." *New Journal of Chemistry* 17 (1993): 661–681.

Walba, David. "A Topological Hierarchy of Molecular Chirality and other Tidbits in Topological Sterochemistry." In *New Developments in Molecular Chirality*. Vol. 5. Edited by P. Mezey, 119–129. Boston: Kluwer Academic Publishers, 1991.

Wolf, Marek. "Applications of Statistical Mechanics in Prime Number Theory," a preprint paper which is summarized by Matthew R. Watkins in his Web page devoted to Marek Wolf. http://www.maths.ex.ac.uk/~mwatkins/zeta/supersymmetry.htm. See also Marek Wolf, "Applications of Statistical Mechanics in number Theory," *Physica A274* (1999): 149–157.

Zarembinski, Thomas I., Youngchang Kim, Kelly Peterson, Dinesh Christendat, Akil Dharamsi, Cheryl H. Arrowsmith, Aled M. Edwards, and Andrzej Joachimiak. "Deep Trefoil Knot Implicated in RNA Binding Found in an Archaebacterial Protein." *Proteins* 50 (2002): 177–183.

Chapter 5

Program Code for Generating the Solenoid

The following pseudocode computes (x, y, z) coordinates for the centers of the nested tubes in the solenoid construction.

- *level: nesting level*
- *circlepts: number of steps around longitudinal circle*
- *zr, zi: longitudinal angle, as a complex number pair*
- *wr, wi: location inside the cross-sectional disk, as a complex number pair.*

```
circlepts = 36;
pi = 3.14159;
for i = 0 to circlepts do
begin
angle = 2 * pi * i / circlepts;
x = cos(angle); (* initial longitudinal *)
```

```
y = sin(angle); (* angular position *)
zr = x; zi = y; (* as a complex number *)
wr = 0; wi = 0; (* cross-section location *)
    for j = 1 to level do
    begin
    wr = wr + zr / 4;
    wi = wi + zi / 4;
    zx = zr * zr-zi * zi; (*complex squaring*)
    zy = 2 * zr * zi; (* of z *)
    zr = zx; zi = zy;
    end;
x = zr * (1 + wr);
y = zi * (1 + wr);
z = wi;
(* The radius of the cross-sectional disk centered at the point (x,y,z) is
1/(2**(level+1)) *)
end;
```

Note: In his 1872 book *Budget of Paradoxes*, Augustus De Morgan explains an equation involving π to an insurance salesman. He is referring to approximation of binomials with the normal distribution, a calculation that requires pi. Robert L. Brown writes the following in the November/December 2002 issue of the American Actualities Association Magazine. *Contingencies* (see http://www.contingencies.org/novdec02/letters.pdf):

For any individual in the group, the chance of surviving (or dying) is binomial. But for the group (of a large number of persons), the Central Limit Theorem says that you can approximate the multiple binomials with the Normal Distribution. And, of course, the evaluation of the Normal requires the constant, pi. I agree . . . that there is no relationship between a circle and the number of people alive at the end of a given time.

Alexander, J. W. "An Example of a Simply Connected Surface Bounding a Region Which Is Not Simply Connected." *Proceedings of the National Academy of Sciences* 10 (1924): 8–10. (On Alexander's horned sphere.)

Biggs, Norman. "The Development of Topology." In *Möbius and His*

Band: Mathematics and Astronomy in Nineteenth-Century Germany. Edited by Fauvel, J., R. Flood, and R. Wilson. Oxford, England: Oxford University Press, 1993.

Boas, Ralph. P., Jr. "Möbius Shorts." *Mathematics Magazine* 68, no. 2 (April 1995): 127.

Bogomolny, Alexander. "Barycentric coordinates." Cut the Knot. http://www.cut-the-knot.org/triangle/barycenter.shtml.

Bouwkamp, C. J. and A. J. W. Duijvestijn. "Catalogue of Simple Perfect Squared Squares of Orders 21 Through 25." Eindhoven University of Technology, Dept. of Math. Report 92-WSK-03, November 1992.

Bouwkamp, C. J. and A. J. W. Duijvestijn. "Album of Simple Perfect Squared Squares of Order 26." Eindhoven University of Technology, Faculty of Mathematics and Computing Science, EUT Report 94-WSK-02, December 1994.

Brooks, R. L., C. A. B. Smith, A. H. Stone, and W. T. Tutte. "The Dissection of Rectangles into Squares." *Duke Mathematics Journal* 7, no. 1 (1940): 312–340.

Browne, Cameron. Cameron's Art Page. http://members.optusnet.com.au/cameronb/art-1.htm. (On Alexander's horned sphere.)

Crowe, Michael. "August Ferdinand Möbius." In *Dictionary of Scientific Biography*. Edited by Charles Gillispie. New York: Charles Scribner, 1974. (This reference contains Möbius's quotation describing rotating a cube in a space of four dimensions.)

Deléglise, Marc, and Joöl Rivat. "Computing the Summation of the Möbius Function." *Experimental Mathematics* 5, no. 4 (1996): 291–295.

Devlin, Keith. "The Mertens Conjecture," *Bulletin of the Irish Math Society* 17 (1986): 29–43.

Earls, Jason. Jason Earls's Store. http://www.lulu.com/JasonEarls.

Fay, Temple. "The Butterfly Curve." *American Mathematical Monthly* 96, no. 5 (1989): 442–443.

Gardner, Martin. *The Colossal Book of Mathematics: Classic Puzzles, Paradoxes, and Problems.* New York: Norton, 2001. (Describes the cannibal torus.)

Gardner, Martin. "The Island of Five Colors." In *Fantasia Mathematica.* Edited by Clifton Fadiman. New York: Simon and Schuster, 1958.

Gardner, Martin. *Hexaflexagons and Other Mathematical Diversions: The First Scientific American Book of Mathematical Puzzles and Games.* 1959. Reprint, Chicago: University of Chicago Press, 1988. (Describes how to turn a torus inside out.)

Gardner, Martin. *Mathematical Magic Show: More Puzzles, Games, Diversions, Illusions and other Mathematical Sleight-of-Mind from Scientific America.* New York: Vintage, 1978.

Gardner, Martin. "Squaring the Square." In *The Second Scientific American Book of Mathematical Puzzles and Diversions.* Reprint, Chicago: University of Chicago Press, 1987.

Gray, Jeremy. "Möbius's Geometrical Mechanics." In *Möbius and his Band: Mathematics and Astronomy in Nineteenth-Century Germany.* Edited by J. Fauvel, R. Flood, and R. Wilson. Oxford, England: Oxford University Press, 1993. (This provides an excellent description of Möbius's barycentric calculus and a figure similar to that in figure 5.33.)

Hart, George W. "The Millennium Bookball." http://www.mi.sanu.ac.yu/vismath/hart/.

Mackenzie, Dana. "What is the name of Euclid is Going on Here?" *Science* 307, no 5714, (March 4, 2005): 1402.

Mandelbrot, Benoit. "A Fractal Life: Interview with Valerie Jamieson." *New Scientist* 184, no. 2473 (November 13, 2004): 50–52.

Möbius, F. A. "Kann von zwei dreiseitigen Pyramiden eine jede in Bezug auf die andere um- und eingeschrieben zugleich heissen?" *Journal für die Reine und Angewandte Mathematik* 3 (1828): 273–278.

Ninham, Barry, and Barry Hughs. "Möbius, Mellin, and Mathematical Physics." *Physica A* 18 (1992): 441–481. (Discusses the Möbius function and other examples of number theory with application to the real world, particularly in the area of physics.)

Odlyzko, A. M. and te Riele, H. J. J. "Disproof of the Mertens Conjecture." *Journal fur die Reine und Angewandte Mathematik* 357 (1985): 138–160.

Pegg, Ed, Jr. "The Möbius Function (and Squarefree Numbers)." Mathematic Association of America Online.
http://www.maa.org/editorial/mathgames/mathgames_11_03_03.html.

Peterson, Ivars. "Surreal Films: A Soapy Solution to the Math Puzzle of Turning a Sphere Inside Out." *Science News* 154, no. 15 (October 10, 1998): 232. http://www.sciencenews.org/pages/sn_arc98/10_10_98/bob1.htm.

Pickover, Clifford. *Surfing Through Hyperspace: Understanding Higher Universe in Six Easy Lessons.* New York: Oxford, University Press 1999.

Pickover, Clifford. *Computers and the Imagination: Visual Adventures Beyond the Edge.* New York: St. Martin's Press, 1991. (Has information on the solenoid.)

Pintz, J. "An Effective Disproof of the Mertens Conjecture." *Astérique* 147/148 (1987): 325–346.

Rucker, Rudy. *The Fourth Dimension: A Guided Tour of the Higher Universes.* Reprint, New York: Houghton Mifflin, 1985.

Schofield, Alfred Taylor. *Another World; or, The Fourth Dimension.* 2nd ed. London: Swan Sonnenschein, 1897.

Smale, Stephen. "Differentiable Dynamical Systems." *Bulletin of the American Math Society* 73 (1967): 748–817. (On attractors and solenoids.)

Stewart, Ian. *Math Hysteria: Fun and Games with Mathematics.* New York: Oxford University Press, 2004. (Contains several references on "squaring the square.")

Stover, Jason, H. "A Rate of Convergence for a Particular Estimate of a Noise-Contaminated Chaotic Time Series." http://www.lisp-p.org/ctfs/. (On solenoids.)

Stølum, Hans-Henrik. "River Meandering as a Self-Organization Process." *Science* 271, no. 5256 (March 22, 1996): 1710–1713. (See also Simon Singh, *Fermat's Enigma,* New York: Anchor, 1998, for more information on pi and river lengths.)

te Riele, H. J. J. "Some Historical and Other Notes About the Mertens Conjecture and Its Recent Disproof." *Nieuw Archief voor Wiskunde* (Vierde Serie) 3, no. 2 (1985): 237–243.

von Sterneck, R. D. "Die Zahlentheoretische Funktion (p/n) bis zur Grenze 500000." *Akad. Wiss. Wien Math.-Natur. Kl. Sitzungsber. IIa,* 121 (1912): 1083–1096.

Weisstein, Eric. MathWorld, a Wolfram Web Resource, s.v. "Möbius Strip." http://mathworld.wolfram.com/MoebiusStrip.html.

Weisstein, Eric. MathWorld, a Wolfram Web Resource, s.v. "Mertens Conjecture." http://mathworld.wolfram.com/MertensConjecture.html.

Weisstein, Eric. MathWorld, a Wolfram Web Resource, s.v. "Möbius Function." http://mathworld.wolfram.com/MoebiusFunction.html.

Weisstein, Eric. MathWorld, a Wolfram Web Resource, s.v. "Möbius Shorts." http://mathworld.wolfram.com/MoebiusShorts.html.

Wikipedia Encyclopedia, s.v. "Homeomorphism." http://en.wikipedia.org/wiki/Homeomorphism.

Wikipedia Encycolpedia, s.v. "Möbius Strip." http://en.wikipedia.org/wiki/M%F6bius_strip.

Wikipedia Encyclopedia, s.v. "Möbius Function (Talk)."
http://en.wikipedia.org/wiki/Talk:M%F6bius_function.

Weisstein, Eric, Margherita Barile, et al. MathWorld, a Wolfram Web
Resource, s.v. "Möbius Tetrahedra."
http://mathworld.wolfram.com/MoebiusTetrahedra.html.

Weisstein, Eric. MathWorld, a Wolfram Web Resource, s.v. "Möbius
Triangles." http://mathworld.wolfram.com/MoebiusTriangles.html.

Chapter 6

Notes:

Visible universe: In 2004, the visible universe was estimated to have a
radius of 78 billion light-years and be 13.7 billion years old. (The radius
of the universe is not 13.7 billion light-years because space continually
expands, with the most rapid expansion occurring early in the birth of
the universe.)

Lines on a sphere: Some scientists do not consider latitude lines on the
Earth as parallel "lines." Except on the equator, latitude "lines" are not
the shortest distance between two points. The great circle path between
points on the Earth, idealized as a sphere, is a geodesic, where the term
"great circle" refers to any circle on a sphere that has the same diameter
as the sphere. In intuitive terms, an elastic band stretched along a path
that is not a geodesic would contract its length for energy reasons to a
nearby shorter path.

How many copies of you exist? Researchers suggest that if the matter and
energy in the universe are created by random quantum fluctuations, as
cosmic inflation dictates, then there will be an infinite number of copies
of the finite configuration of matter and energy in our visible universe,
which is easily contained in a sphere 100 billion light years in diameter.
Scientists believe that a lump of matter and energy enclosed in a finite
sphere can be arranged in only a finite number of ways—due to a restriction known as the "holographic bound." Learn more in Charles Seife's
"Physics Enters the Twilight Zone."

ALGORITHM: How to Create a Banchoff Klein Bottle

```
for(u= 0; u < 6.28; u = u + .2){
for(v= 0; v < 6.28; v = v + .05 {
    x = cos(u)*(sqrt(2) + cos(u/2)*cos(v) + sin(u/2) * sin(v) * cos(v));
    y = sin(u)*(sqrt(2) + cos(u/2)*cos(v) + sin(u/2) * sin(v)*cos(v));
    z = -sin(u/2)*cos(v) + cos(u/2)*sin(v)*cos(v);
    DrawSphereCenteredAt(x,y,x)
    }
}
```

Adams, Colin, and Joey Shapiro. "The Shape of the Universe: Ten Possibilities." *American Scientist* (September/October, 2001): 443–453.

Adams, Fred, and Greg Laughlin. *The Five Ages of the Universe: Inside the Physics of Eternity.* 202–203; Lee Smolin, 1997, *Life of the Cosmos* New York: Oxford University Press: 1997). (Roger Penrose and Stephen Hawking have suggested that the expanding universe is described by the same equations as a collapsing black hole, but with the opposite direction of time. Black holes may be the seeds for other universes. According to John Gribbin, in *Stardust,* the number of baby universes may be proportional to the volume of the parent universe.)

Banchoff, Thomas. *Beyond the Third Dimension: Geometry, Computer Graphics, and Higher Dimensions.* 2nd ed. New York: Freeman, 1996.

Cowen, Ron. "Cosmologists in Flatland: Searching for the Missing Energy." *Science News* 153, no. 9 (February 28, 1998): 139–141.

Davies, Paul. "A Brief History of the Multiverse." *New York Times,* April 12, 2003, late edition final, sec. A. (See also the Nick Bostrom Web page, www.simulation-argument.com.)

Egan, Greg. *Permutation City.* New York: HarperCollins, 1994.

Klarreich, Erica. "The Shape of Space." *Science News* 164, no. 19 (November 8, 2003): 296–297.

Moore, Alan, and Kevin O'Neill. "The New Traveller's Almanac." In *League of Extraordinary Gentlemen,* vol. 2. New York: DC Comics, 2003.

Overbye, Dennis. "Universe as Doughnut: New Data, New Debate." *New York Times,* March 11, 2003, late edition final, sec. F.

Peterson, Ivars. "Circle in the Sky: Detecting the Shape of the Universe." *Science News* 153, no. 8 (February, 21, 1998): 123–135.

Pickover, Clifford. *The Paradox of God and the Science of Omniscience.* New York: St. Martin's Press/Palgrave, 2001.

Pickover, Clifford. *Surfing Through Hyperspace: Understanding Higher Universes in Six Easy Lessons.* New York: Oxford University Press, 1999.

Pickover, Clifford. *Time: A Traveler's Guide.* New York: Oxford University Press, 1998.

Rees, Martin. "In the Matrix." Edge. http://www.edge.org/3rd_culture/rees03/rees_p2.html.

Rees, Martin. "Living in a Multiverse." In *The Far Future Universe.* Edited by George Ellis, 65–88. West Conshohocken, Pennsylvania: Templeton Press, 2002.) (Also see the Nick Bostrum's Web site, www.simulation-argument.com.)

Rucker, Rudy. *Seek!* New York: Four Walls Eight Windows, 1999. 150–151. (See the chapter "Goodbye Big Bang," which discusses Andre Linde's baby universes.)

Seife, Charles. "Big Bang's New Rival Debuts with a Splash," *Science* 292, 189–191, (April 13, 2001): 5515.

Seife, Charles. "Physics Enters the Twilight Zone." *Science* 305, no. 5683 (July 23, 2004): 464–466. (This article discusses cosmic inflation and the size of our universe, quoting physicist Max Tegmark who says, "Inflation generically predicts infinite space. Not just big, but infinite." See also http://www.sciencemag.org/cgi/content/full/305/5683/464.)

Seife, Charles. "Polyhedral Model Gives the Universe an Unexpected Twist." *Science* 302 (October 10, 2003): 209.

Stoll, Cliff. "Drinking Mug Klein Bottles–for the Thirsty Topologist." Acme Klein Bottle. http://www.kleinbottle.com/drinking_mug_klein_bottle.htm.

Stoll, Cliff, "Acme Klein Bottle," http://www.kleinbottle.com/meter_tall_klein_bottle.html.

Wells, David. *The Penguin Dictionary of Curious and Interesting Geometry.* New York: Penguin Books, 1992.

Wright, Edward, L. "How can the Universe be Infinite if It Was All Concentrated into a Point at the Big Bang?" http://www.astro.ucla.edu/~wright/infpoint.html.

Chapter 7
Albert, Don. "Möbius heart." http://home.earthlink.net/~donaldwalbert/Pages/DAGAdesignPrint.html.

Boittin, Margaret, Erin Callahan, David Goldberg, and Jacob Remes, (Yale University), "Math that Makes You Go Wow." http://www.math.ohio-state.edu/~fiedorow/math655/yale/random.htm.

Gardner, Martin. *Mathematical Magic Show: More Puzzles, Games, Diversions, Illusions and other Mathematical Sleight-of-Mind* for *Scientific America.* New York: Vintage, 1978.

Key Curriculum Press. "Torus and Klein Bottle Games." http://www.geometrygames.org/TorusGames/.

Krasek, Teja. Teja's S. http://tejakrasek.tripod.com/.

Krawczyk, Robert J., and Jolly Thulaseedas. "Möbius Concepts in Architecture." http://www.iit.edu/~krawczyk/jtbrdg03.pdf.

Lipson, Andrew. Andrew Lipson's LEGO® Page. http://www.lipsons.pwp.blueyonder.co.uk/.

Miller, George. "Moby Maze" The Puzzle Palace catalogue. http://www.puzzlepalace.com/puzzle.php?catalogNum=200405.

Miller, Jeff. Images of Mathematicians on Postage Stamps. http://jeff560. tripod.com/.

Wikipedia Encyclopedia, s.v. "Penrose tiling." http://en.wikipedia.org/wiki/Penrose_tiling.

Peterson, Ivars. "Möbius at Fermilab." *Science News* 158, no. 10 (September 2, 2000). http://www.sciencenews.org/articles/20000902/math trek.asp.

Pickover, Clifford. *The Zen of Magic Squares, Circles, and Stars: An Exhibition of Surprising Structures across Dimensions.* Princeton, New Jersey: Princeton University Press, 2002. (This book has many examples of knight's tours.)

Rogger, Andre. "Away with the Alps, Open Up the View to the Mediterranean." DB Artmag. http://www.deutsche-bank-kunst.com/art/2003/15/e/1/153.php. (Discusses Max Bill's Kontinuität, 1986. For a more general discussion, see "Deutsche Bank's Art," http://www.deutsche-bank-kunst.com/beta30/english/ie1024/100xkunst/100xkunst_1986.htm.)

Scharein, Robert. "Complex Knots" and "Möbius Strip Knots." http://www.cecm.sfu.ca/~scharein/KnotPlot/complex/. (See also: http://www.cecm.sfu.ca/~scharein/projects/moebius/ and http://hypnagogic.net/

Shoulson, Mark E. "Möbius strip." Meson.org. http://web.meson.org/topology/mobius.html.

Stewart, Ian. *Another Fine Math You've Got Me Into.* New York: Freeman, 1992. (Describes the knight's tour on various kinds of chessboards.)

Tanton, James. *Solve This: Math Activities for Students and Clubs.* Washington, D.C.: The Mathematical Association of America, 2001.

Watkins, John J. *Across the Board: The Mathematics of Chessboard Problems.* Princeton, New Jersey: Princeton University Press, 2004.

Chapter 8

Note: Several manuscript readers asked me what the word "Viconian" means in John Barth's foreword to *Lost in the Funhouse*. This word refers to the theories of the Neapolitan philosopher, Giambattista (Giovanni Battista) Vico (1668–1744), which concerned the cyclical nature of culture. In particular, Vico asserted that history is cyclical and that each cycle consists of three different ages–an age of gods, an age of heroes, and an age of humans. These are followed by a short transitional age that initiates the next cycle.

Barth, John. "Art of the Story: Interview with John Barth." By Elizabeth Farnsworth. *NewsHour with Jim Lehrer* (November 18, 1998). http://www.pbs.org/newshour/bb/entertainment/july-dec98/barth_11-18.html.

Kasman, Alex. "Geometry, Trigonometry, and Topology in Math Fiction." *Mathematical Fiction: A List Compiled by Alex Kasman, College of Charleston.* http://math.cofc.edu/faculty/kasman/MATHFICT/search.php?orderby=title&go=yes&topics=gtt.

Pickover, Clifford. *Sex, Drugs, Einstein, and Elves: Sushi, Psychedelics, Parallel Universes, and the Quest for Transcendence.* Petaluma, California: Smart Publications, 2005.

Sorrentino, Christopher. "Reading Coleman Dowell's *Island People.*" Center for Book Culture. http://centerforbookculture.org/context/no3/sorrentino.html.

Conclusion

Boittin, Margaret, Erin Callahan, David Goldberg, and Jacob Remes. "Math that Makes You Go Wow." http://www.math.ohio-state.edu/~fiedorow/math655/yale/random.htm.

Gardner, Martin. *Order and Surprise.* Buffalo, New York: Prometheus Books, 1983. (Readers should look at the chapter "The Adventures of Stanislaw Ulam.")

Mandelbrot, Benoit. "A Fractal Life: Interview with Valerie Jamieson." *New Scientist* 184, no. 2473 (November 13, 2004): 50–52.

Pickover, Clifford. *Wonders of Numbers: Adventures in Mathematics, Mind, and Meaning.* New York: Oxford University Press, 2001. (Contains additional examples of amateur mathematicians making significant discoveries.)

Wallace, Jonathan. "Proust's Ruined Mirror." *The Ethical Spectacle* 5, no. 2 (February, 1999). http://www.spectacle.org/299/main.html.

General Reading
Alexander, Neil. "Magic tricks." Conjuror. http://www.conjuror.com/. (Teaches how to perform the Afghan bands trick.)

Ball, W. W. R. and H. S. M. Coxeter. *Mathematical Recreations and Essays.* 13th ed. New York: Dover, 1987. (127–128)

Bogomolny, A. "Möbius Strip." Interactive Mathematics Miscellany and Puzzles. http://www.cut-the-knot.org/do_you_know/moebius.shtml.

Boittin, Margaret, Erin Callahan, David Goldberg, and Jacob Remes. "Math that Makes You Go Wow." http://www.math.ohio-state.edu/~fiedorow/math655/yale/random.htm

Bondy, John Adrian and U. S. R. Murty. *Graph Theory with Applications.* New York: North Holland, 1976. (243)

Bool, F. H., J. R. Kist, J. L. Locher, and F. Wierda. *M. C. Escher: His Life and Complete Graphic Work.* New York: Abrams, 1982.

Crowe, Michael. "August Ferdinand Möbius." In *Dictionary of Scientific Biography.* Edited by Charles Gillispie. (New York: Charles Scribner, 1974.

Derbyshire, J. *Prime Obsession: Bernhard Riemann and the Greatest Unsolved Problem in Mathematics.* New York: Penguin, 2004.

Fauvel, John, Raymond Flood, and Robin Wilson. *Möbius and His Band: Mathematics and Astronomy in Nineteenth-Century Germany.* New York: Oxford University Press, 1993.

Gardner, Martin. *Mathematics, Magic and Mystery.* New York: Dover, 1956.

Gardner, Martin. *Mathematical Magic Show: More Puzzles, Games, Diversions, Illusions and other Mathematical Sleight-of-Mind* for *Scientific America.* New York: Vintage, 1978.

Gardner, Martin. *The Sixth Book of Mathematical Games from Scientific American.* Chicago: University of Chicago Press, 1984. (10)

Geometry Center. "The Möbius Band." http://www.geom.umn.edu/zoo/features/möbius/.

Gray, Alfred. "The Möbius Strip," Chapter 14 in *Modern Differential Geometry of Curves and Surfaces with Mathematica.* 2nd ed. Boca Raton, Florida: CRC Press, 1997. (325–326)

Hunter, J. A. H., and J. S. Madachy. *Mathematical Diversions.* New York: Dover, 1975. 41–45

Isaksen, D. C., and A. P. Petrofsky. "Möbius knitting." In *Bridges: Mathematical Connections in Art, Music, and Science Conference Proceedings.* Edited by R. Sarhangi. 1999. Winfield, Kansas: Southwestern College Bridges (See http://www.sckans.edu/~bridges/.)

Kasman, Alex. Math Fiction. A List Compiled by Alex Casman, College of Charleston. http://math.cofc.edu/faculty/kasman/MATHFICT.

Madachy, Joseph S. *Madachy's Mathematical Recreations.* New York: Dover, 1979. (7)

M. C. Escher Foundation. M. C. Escher: The Official Website. http://www.mcescher.com.

Pappas, Theoni. *The Joy of Mathematics.* San Carlos, California: Wide World Publishing/Tetra, 1989.

Möbius, August. *Gesammelte Werke.* Edited by Richard Baltzer, Felix Klein, and Wilhelm Scheibner. 4 vols. Leipzig: reprinted, Dr. Martin Sändig oHg, Wisbaden, 1967.

O'Connor, John J., and Edmund F. Robertson. August Ferdinand Möbius. (Biographical sketch), http://www-history.mcs.st-andrews. ac.uk/history/Mathematicians/Mobius.html.

Peterson, Ivars. "Möbius in the Playground." Ivars Peterson's Math Trek. *Science News Online* (May 22,1999). http://www.sciencenews.org/sn_arc99/5_22_99/mathland.htm.

Peterson, Ivars. "More than Just a Plane Game." Ivars Peterson's Math Trek. *Science News Online* (March 14, 1998). http://www.sciencenews.org/sn_arc98/3_14_98/mathland.htm.

Peterson, Ivars. "Recycling topology." Ivars Peterson's Math Trek. *Science News Online* (September 28, 1996). http://www.sciencenews.org/sn_arch/9_28_96/mathland.htm.

Peterson, Ivars. "Möbius and his Band." *Science News Online* 158, no. 2 (July 8, 2000). http://www.sciencenews.org/articles/20000708/mathtrek.asp.

Weisstein, Eric W. MathWorld, a Wolfram Web Resource, s.v. "Polyhedral Formula." http://mathworld.wolfram.com/PolyhedralFormula.html. (Provides a definition of Euler's polyhedron formula.)

Wells, David. *The Penguin Dictionary of Curious and Interesting Geometry.* London: Penguin, 1991.

Wikipedia Encylopedia. s.v. "Möbius Strip." http://en.wikipedia.org/wiki/M%F6bius_strip.

ABOUT THE AUTHOR

Clifford A. Pickover received his Ph.D. from Yale University's Department of Molecular Biophysics and Biochemistry. He graduated first in his class from Franklin and Marshall College, after completing the four-year undergraduate program in three years. His many books have been translated into Italian, French, Greek, German, Japanese, Chinese, Korean, Portuguese, Spanish, Turkish, and Polish. One of the most prolific and eclectic writers of our time, Pickover is author of the popular books: *A Passion for Mathematics* (Wiley, 2005), *Sex, Drugs, Einstein, and Elves* (Smart Publications, 2005), *Calculus and Pizza* (Wiley, 2003), *The Paradox of God and the Science of Omniscience* (Palgrave/St. Martin's Press, 2002), *The Stars of Heaven* (Oxford University Press, 2001), *The Zen of Magic Squares, Circles, and Stars* (Princeton University Press, 2001), *Dreaming the Future* (Prometheus, 2001), *Wonders of Numbers* (Oxford University Press, 2000), *The Girl Who Gave Birth to Rabbits* (Prometheus, 2000), *Surfing Through Hyperspace* (Oxford University Press, 1999), *The Science of Aliens* (Basic Books, 1998), *Time: A Traveler's Guide* (Oxford University Press, 1998), *Strange Brains and Genius: The Secret Lives of Eccentric Scientists and Madmen* (Plenum, 1998), *The Alien IQ Test* (Basic Books, 1997), *The Loom of God* (Plenum, 1997), *Black Holes-A Traveler's Guide* (Wiley, 1996), and *Keys to Infinity* (Wiley, 1995). He is also the author of numerous other highly acclaimed books, including *Chaos in Wonderland: Visual Adventures in a Fractal World* (1994), *Mazes for the Mind: Computers and the Unexpected* (1992), *Computers and the Imagination* (1991), and *Computers, Pattern, Chaos, and Beauty* (1990), all published by St. Martin's Press. He has published more than two hundred articles concerning topics in science, art, and mathematics. He is also coauthor, with Piers Anthony, of *Spider Legs*, a novel once listed as the second-best-selling science fiction title on Barnes & Noble.com. Pickover is currently an associate editor for the scientific journal *Computers and Graphics* and is an editorial board member for *Odyssey, Leonardo,* and *YLEM.*

Editor of the books *Chaos and Fractals: A Computer Graphical Journey* (Elsevier, 1998), *The Pattern Book: Fractals, Art, and Nature* (World Scientific, 1995),

Visions of the Future: Art, Technology, and Computing in the Next Century (St. Martin's Press, 1993), *Future Health* (St. Martin's Press, 1995), *Fractal Horizons* (St. Martin's Press, 1996), and *Visualizing Biological Information* (World Scientific, 1995), and coeditor of the books *Spiral Symmetry* (World Scientific, 1992) and *Frontiers in Scientific Visualization* (Wiley, 1994), Dr. Pickover is interested in finding new ways to continually expand creativity by melding art, science, mathematics, and other seemingly disparate areas of human endeavor. He is the author of the popular "Neoreality" science-fiction series (*Liquid Earth, Sushi Never Sleeps, The Lobotomy Club*, and *Egg Drop Soup*), in which characters explore strange realities.

The *Los Angeles Times* recently proclaimed, "Pickover has published nearly a book a year in which he stretches the limits of computers, art and thought." Pickover received first prize in the "Beauty of Physics Photographic Competition," sponsored by the Institute of Physics. His computer graphics have been featured on the covers of many popular magazines, and his research has recently received considerable attention by the press—including *CNN*'s "Science and Technology Week," *The Discovery Channel, Science News, The Washington Post, Wired,* and *The Christian Science Monitor*—and also in international exhibitions and museums. *OMNI* magazine described him as "Van Leeuwenhoek's twentieth century equivalent." *Scientific American* featured his graphic work several times, calling it "strange and beautiful, stunningly realistic." *Wired* magazine wrote, "Bucky Fuller thought big, Arthur C. Clarke thinks big, but Cliff Pickover outdoes them both." Pickover holds over thirty U.S. patents, mostly concerned with novel features for computers.

Dr. Pickover is currently a research staff member at the IBM T. J. Watson Research Center, where he has received forty invention achievement awards and three research division awards. For many years, Dr. Pickover wrote *Discover* magazine's Brain-Boggler column, and he currently writes the Brain-Strain column for *Odyssey*. His puzzle calendars and cards are designed for both children and adults and have sold hundreds of thousands of copies.

Dr. Pickover's hobbies include the practice of Ch'ang-Shih Tai-Chi Ch'uan and Shaolin Kung Fu, raising golden and green severums (large Amazonian fish), and piano playing (mostly jazz). He is also a member of the SETI League, a group of signal processing enthusiasts who systematically search the sky for intelligent extraterrestrial life. Visit his Web site, which has received over a million visits: http://www.pickover.com. He can be reached at P.O. Box 549, Millwood, New York 10546-0549, USA.

INDEX